WEED, WEEDICIDES AND WEED CONTROL PRINCIPAL AND PRACTICE

Weed, Weedicides and Weed Control

Principal and Practice

R. C. Mandal
Ex. Director-in-charge
Tuber & Cashew
Indian Council of Agricultural Research

Forwarded by
Dr. B. P. Pal
Ex. Director General
Indian Council of Agricultural Research
New Delhi

AGROBIOS (INDIA)

Published by:
Agrobios (India)
Behind Nasrani Cinema
Chopasani Road, Jodhpur
Ph.: 0291-2643993, 2642319
Email: agrobiosindia@gmail.com

AGROBIOS (INDIA)

AGROBIOS INDIA (2026)

ISBN (13): 978-81-7754-021-1

Published by: Dr. Updesh Purohit for Agrobios (India), Jodhpur
Laser typeset at: Yashee Computers, Jodhpur
Cover Design by: Reena
Printed at: Agrobios Digitals, Jodhpur

Foreword

To meet the food needs of the rapidly growing population, the farmers have to achieve economic efficiency by intensifying their farming system. Weed science has assumed great importance and weed control technology is considered as a part of the production system. Innumerable weed species grow under various environments competing successfully with neighbouring crops and hence an integrated approach to protect the crop from enormous losses by weeds is needed, particularly by changing the system from traditional manual to modern herbicidal weed control.

The author has briefly described monocotyledonous, dicotyledonous, parasitic and aquatic weeds, their allelopathy and also suitable herbicidal treatments for control of weeds occurring in annual, biennial and perennial crops. Useful illustrations have also been provided. Every attempt has been made to provide sufficient information in each chapter in handling particular weed problems. Many of these practical weed control techniques have proved of great value. The rapidly increasing specialization of weed control techniques has made it desirable to emphasize the broadly applicable principles and basic methods and the scientific basis underlying control measures.

The author has dealt with the subject in a commendable manner and I am confident that the present book will be very useful to teachers in colleges and vocational training centres and also to research workers, students and extension specialists. The book will, I hope be welcomed as making a substantial contribution towards filling the large gap in the available weed science literature. I sincerely wish that this kind of practical guide should reach all those who are interested in the practical aspects of the important subject of weed control.

B. P. Pal
Ex-Director General
Indian Council of Agricultural Research
New Delhi

Preface

Weeds pose serious problems in both crop and non-cropped areas and water-bodes. They infest lawn and gardens; become troublesome along highways, railroads and industrial areas; in irrigation and drainage systems; and create health hazards to animal and human beings.

Though a handbook of 'some South Indian weeds' was published in 1932, the weed control as a Science was initiated since 1942. Since then the primitive methods of weed control are blended with the newest herbicide control techniques and 2,4-D was first used on a large scale in 1947. With the growing interest in weeds and weed control by the scientists of different disciplines, namely Ecologist, Plant Physiologist, Agronomist, Biochemist and others, the weed science has assumed great importance. It has made a considerable progress during last 40 years and progressively assuming an independent and significant branch of biological research. Workers in this field are now employed by educational institutions, government agencies, herbicide manufacturers and others. Chemical companies are spending millions of rupees in synthesizing, screening and developing new herbicides with trade names. With the availability of effective herbicides, there is an urgent need to bring together the present concept and modern techniques of weed control, along with complete enumeration of weed problems; so as to cope with innumerable weed species, grown under various environments. However, since conditions vary as to methods of application and environmental conditions, the dose may be adjusted according to the need and to be applied judiciously, based on the instructions printed on the label of the herbicides.

The author has attempted to present the subject in an useful way and readable manner so as to understand easily. It will be helpful to research workers, college and University teachers, Vocational teachers, extension specialists, herbicide sale representatives and progressive farmers. Weed control practices suggested are based on research reports of several workers in this field and his own experience. However, farmers are not quite familiar to the use of herbicides as is the case of pesticides and fungicides; only because they do not have adequate knowledge in the techniques of application, availability and also cost involvement.

The book contains classification and description of monocot, dicot, aquatic and parasitic weeds; reproduction and dissemination of weeds, allelopathy, weed control methods, classification of herbicides and mode of herbicidal action; selective weed control in field crops; plantations and Horticultural crops, grass lands and ornamentals; residual effect of herbicides, along with useful illustrations.

A couple of vary good text books on weed science from India and abroad are available which are being used as guides to teach weed science courses. However, this book is a practical guide, provided with up-to-date information in a comprehensive and understandable manner, dealing with all the practical aspects.

It is a great pleasure to express my sincere thanks to Dr. R. D. Iyer, Head of Genetics Division, Central Plantation Crops Research Institute, Kasaragod who took keen interest in reading the chapters to make it technically correct. My sincere gratitude is due to all the Indian and foreign authors and weed science workers whose contributions, have been reflected in this volume. Due acknowledgements have been given to those works and workers at the appropriate place.

R. C. Mandal

Contents

1. DEFINITION AND SOME BASIC CONCEPTS 1

1. Introduction 1
   - *1.1. Basic Concept 1*
   - *1.2. Losses Caused by Weeds 2*
2. Effect of Weeds and their Competition 2
   - *2.1. Competition of Weeds for Water, Light and Mineral Nutrients 2*
   - *2.2. Competition of Weeds for Space 4*
   - *2.3. Weeds Increase Crop Pests 4*
   - *2.4. Weeds Interfere with Crop Culture 4*
   - *2.5. Weeds Reduce Crop Quality 4*
   - *2.6. Weeds Harm Animal Health 5*
   - *2.7. Weeds Harm Human Health 5*
   - *2.8. Allelopathic Effects of Weeds 5*
   - *2.9. Effect of Parasitic Weeds 5*
   - *2.10. Weeds Contaminate Water Body 6*
   - *2.11. Weeds Interfere in Non-Cropped Lands 6*
3. Classification of Weeds 6
   - *3.1. According to Ontogeny of Weeds 6*
   - *3.2. According to Cotyledon Characters 7*
   - *3.3. According to Nature of Stem 7*
   - *3.4. According to Soil 8*
   - *3.5. According to Climate 8*
   - *3.6. According to Association 8*
   - *3.7. According to Situation 8*
4. Ecology of Weeds 8
   - *4.1. Weed Ecology 8*
   - *4.2. Survival Mechanism of Weeds 9*
5. Dispersal of Weeds 10
   - *5.1. Wind as an Agent 10*
   - *5.2. Water as an Agent 10*
   - *5.3. Animal and Man as Agents 10*
   - *5.4. Farm Machineries Carry Weed Seeds 11*

5.5. Manures Carry Weed Seeds *11*
6. Prevention, Eradication and Control of Weeds 11
6.1. Prevention of Weeds *11*
6.2. Eradication of Weeds *11*
6.3. Weed Control Methods *12*
7. Formulation of Weedicides and Calculations 14
7.1. Formulations *14*
7.2. Calculations *15*
8. Weedicides Application Equipments 16
8.1. Spraying Method and Spray Volume *16*
8.2. Parts of the Sprayer and their Functions *16*
8.3. Cleaning the Sprayer *17*
9. Safety Precautions for Handing Herbicides and their Applications 18
First-Aid and Antidotes *19*
10. Absorption and Translocation of Herbicides 19
10.1. Physiological Response to Herbicides *19*
10.2. Morphological Responses to Herbicides *19*
10.3. Absorption and Translocation *20*
11. Effect of Herbicides Applied on Soil 21
11.1. Persistence in the Soil *22*

2. MONOCOT WEEDS 24
Grasses – Gramineae 24
1. Agropyron repens *24*
2. Andropogon annulatus *26*
3. Avena fatua *26*
4. Brachiaria remosa *27*
5. Cenchrus catharticus *28*
6. Chloris barbata *28*
7. Cynodon dactylon *28*
8. Digitaria sanguinalis *30*
9.(a) Echinochloa colonum *30*
9. (b) Echinochloa crusgalli *32*
10. Eleusine indica *33*
11. Eragrostis pilosa *33*
12. Lolium temulentum *33*
13. Oryza sativa var. fatua *34*
14. Panicum repens *34*
15. Paspalum spp. *35*
16. Phalaris minor *35*
17. Poa annua *36*

18. Saccharum spontaneum *36*
19. Setaria glauca *37*
20. Sorghum halepense *38*
B. Sedges – Cyperaceae 39
21. Cyperus esculentus *39*
22. Cyperus rotundus *39*
23. Fimbristylis annua *41*
C. Liliaceae 42
24. Allium spp. *42*

3. DICOT WEEDS 46
1. *Acalypha indica* 46
2. *Acanthosperum hispidum* 49
3. *Achyranthus aspera* 51
4. *Amaranthus spinosus* 51
5. *Amaranthus viridis* 52
6. *Anagalis arvensis* 53
7. *Argemone mexicana* 54
8. *Asteracantha longifolia* 55
9. *Boerhaavia diffusa* 55
10. *Calotrpis gigantean* 56
11. *Carthamus oxycantha* 56
12. *Cassia occidentalis* 57
13. *Cassia tora* 57
14. *Celosia argentea* 58
15. *Chenopodium album* 59
16. *Cirsium arvense* 59
17. *Cleome viscos* 60
18. *Convolvulus arvensis* 61
19. *Corchorus acutangulus* 63
20. *Cucumis trigonus* 64
21. *Datura stramonium* 64
22. *Daucus carota* 65
23. *Digera arvensis* 65
24. *Eupatorium odoratum* 66
25. *Euphorbia hirta* 67
26. *Gynandropsis pentaphylla* 68
27. *Indigofera glandulosa* 68
28. *Ipomoea purpurea* 68
29. *Lantana camara* 69
30. *Lathyrus aphaca* 70
31. *Leucas aspera* 71

32. *Melilotus alba* Sym. *vulgaris* 71
33. *Meliolotus indica* Syn. *parviflora* 72
34. *Mimosa pudica* 73
35. *Mollugo verticillata* 73
36. *Opuntia dillenii* 76
37. *Oxalis acetosella* 76
38. *Parthenium hysterophorus* 77
39. *Phyllanthus niruri* 79
40. *Physalis minima* 79
41. *Pluchea lanceolata* 80
42. *Polygonum plebejum* 80
43. *Portulaca oleracea* 81
44. *Solanum xanthocarpum* 81
45. *Sonchus* spp. 82
46. *Spergula arvensis* 82
47. *Stellaria media* 83
48. *Trianthema monogyna* syn. *portulacastrum* 83
49. *Tribulus terrestris* 84

4. PARASITIC WEEDS AND THEIR CONTROL 89
1. *Cuscuta* spp 89
2. *Loranthus* spp 91
3. *Orobanche aegyptiaca* 91
4. *Striga asiatica* 92

5. AQUATIC WEEDS 94
1. Free-Floating Species 96
2.Rooted Floating-Leaf Species 96
3. Sub-Mersed Species 97
Algae *97*
Methods of Aquatic Weed Control 97
A. Mechanical/Physical Control *97*
B. Chemical Control *98*
C. Biological Control *100*
Aquatic Weeds 101
1. Algae *101*
2. Alisma plantago *102*
3. Alternanthera phylloxeroides *102*
4. Azolla imbricate *103*
5. Commelina bengallensis *103*
6. Cyperus difformis *104*
7. Eichhornia crassipes *104*
8. Eleocharis plantaginea *105*

9. Elodea canadensis 106
10. Heteranthera limosa 106
11. Hydrilla verticillata 106
12. Hydrocotyle umbellate 107
13. Ipomoea aquatica 108
14. Juncos effuses 108
15. Jussiea repens 108
16. (a) Lemma Minor (b) Spirodela polyrhiza 109
17. Marsilea minuta 109
18. Monochoria vaginalis 110
19. Myriohpyllum spicatum 110
20. Najas minor 111
21. Nelumbo nucifer 111
22. Nuphar lutea 111
23. Nymphaca alba 112
24. Panicum purpurascens 112
25. Paspalum fluitans 113
26. Pharagmites communis 114
27. Pistia stratiotes 114
28. Pontederia cordata 115
29. Potamogeton spp. 115
30. Sagittaria guayanensis 116
31. Salvinia spp. 116
32. Scirpus articulatus 117
33. Typha spp. 117
34. Utricularia vulgaris 118
35. Vallisneria spiralis 118

6. ALLELOPATHY IN RELATION TO WEEDS 123
Introduction 123
A. Interference of Monocot Weeds on Crops 123
1. Agropyron repens (Quack Grass) 123
2. Avena fatua (Wild Oat) 124
3. Cynodon dactylon (Bermuda Grass) 124
4. Cyperus esculentus (Yellow Nut Sedge) 124
5. Cyperus rotundus (Purple Nutsedge) 124
6. Digitaria sanguinalis (Crab Grass) 125
7. Echinochloa crusgalli (Barnyard Grass) 125
8. Eleusine indica (Goose Grass) 125
9. Lolium multiflorum (Italian Rye Grass) 125
10. Setaria glauca (Yellow Foxtail) and S. faberii (Giant Foxtail) 125

11. Sorghum halepense (Johnson Grass) *126*
B. Interference of Dicot Weeds on Crops 126
1. Amaranthus dubius (Pigweed) *126*
2. Ambrosia artemistifolia (Commno Ragweed) *126*
3. Boerhaavia diffusa (Spreading Spiderling) *126*
4. Celosia argentea (Cock's Comb) *126*
5. Chenopodium album (Lambs Quarters) *127*
6. Cirsium arvense (Canada Thistle) *127*
7. Cucumis trigonus (Wild Cucumber) *127*
8. Convolvulus arvensis (Field Bindweed) *127*
9. Digera arvensis (Digera) *127*
10. Euphorbia spp. (Spurge) *127*
11. Melilotus alba (White Sweet Clover) *128*
12. Partheium hysterophorus (Ragweed, Congress Grass) *128*
13. Phalaris spp. (Canary Grass) *128*
14. Polygonum spp. (Knotweed) *128*
15. Portulaca oleracea (Purselane) *128*
16. Rhus copalina *128*
17. Salsola kali (Russian Thistle) *129*
18. Sonchus arvensis (Sowthistle) *129*

7. WEED CONTROL IN CROPS 130
Selection of Crop Varieties for Tolerance of Herbicides 130
A. Weed Control in Cereals 131
1. Rice (Oryza sativa) Gramineae *131*
2. Maize (Zea mays) Graminiae *133*
3. Fodder Maize and Teosinte (Euchlaena mexicana) *134*
4. Sorghum (Jowar): (Sorghum vulgare) Gramineae *134*
5. Fodder Jowar and Bajra *135*
6. Pearl millet (Bajra): (Pennisetum typhoides) Gramineae *135*
7. Finger Millet (Ragl): (Elusine coracana) – Gramineae *136*
8. Wheat (Triticum vulgare) and Barley (Hordium vulgare) – Gramineae *136*
B. Weed Control in Pulses Legumes 136
1. Pigeonpea (Cajanas cajan) – Papilionaceae *137*
2. Cowpea (Vigna sinensis) – Papilionaceae *137*
3. Black Gram (Phaseolus mungo) and Green Gram

(Phaseolus aureus) – Papilionaceae *137*
4. Broad-Bean (Dolichos lablab) – Papilionaceae *138*
5. Cluster Bean (Cyamopsis tetragonoloba) – Papilionaceae *138*
6. Peas (Pisum sativum) – Papilionaceae *138*
7. Chickpea (Bengal Gram) (Cicer arietinum) – Papilionaceae *139*
C. Weed Control in Oilseed Crops 139
1. Groundnut (Peanut) (Arachis hypogaea) – Papilionaceae *139*
2. Soybean (Glycine max) – Papilionaceae *140*
3. Linseed (Linum usitatissimum) – Linaceae *141*
4. Rape and Mustard (Brassica spp.) – Crucifereae *141*
5. Castor Bean (Ricinus communis) Euphorbiaceae and Til (Sesamum indicum) – Pedaliaceae *142*
6. Safflower (Carthamas tinctorius) – Compositae *142*
7. Sunflower (Helianthus annus) – Compositae *142*
d. Weed Control in Fiber Crops 142
1. Cotton (Gossypium spp.) – Malvaceae *142*
2. Jute (Corchorus olitorius & C. capsularis) – Tiliaceae *144*
e. Weed Control in Sugar Crops 144
1. Sugarcane (Saccharum officinarum) – Gramineae *144*
2. Sugarbeet (Beta vulgaris) – Chinopodiaceae *145*
f. Weed Control in Tobacco (*Nicotiana tabacum*) – Solanaceae 146
G. Weed Control in Vegetable Crops 146
1. Potato (Solanum tuberosum) – Solanaceae *147*
2. Tomato (Lycopersicum esculcutum) – Solanaceae *148*
3. Brinjal (Solanum melongena) – Solanaceae *148*
4. Chilies (Capsicum annuum) – Solanaceae *149*
5. Bhendi (Abelmoscus esculentus) – Malvaceae *149*
6. Cabbage, Cauliflower (Brassica oleracea var. capitata) – Cruciferae *149*
7. Cucurbits: Cucumber (Cucumis sativus), Muskmelon (Cucumis melo) Watermelon (Citrullus vulgaris), Pumpkin (Cucurbita pepo) and Squash (Cucurbita maxima) *150*
8. Beans (Vicia faba) and Greenpeas (Pisum sativum) Papilionaceae *150*
9. Leafy Vegetables: Spinach (Spinacea oleracea)

Chenopodiaceae; and Lettuce (Lactuca sativa) – Cruciferae *150*
10. Onion (Allium cepa) and Garlic (Allium sativum) – Liliaceae *151*
H. Weed Control in Root and Tuber Crops 151
1. Radish (Raphanus sativus), Turnip (Brassica napus) Cruciferae and Carrot (Daucus carota) – Umbelliferae *151*
2. Sweet Potato (Ipomoea batatas) – Convolvulaceae *152*
3. Cassava (Tapioca): (Manihot utilissima) – Euphorbiaceae *152*
4. Colocasia and Xanthosoma spp. – Araceae *152*
5. Yams (Dioscorea esculenta, D. alata) – Dioscoreaceae *152*
I. Weed Control in Annual Spices 152
1. Turmeric (Curcuma domestica) and Ginger (Zingiber officinale) – Zingiberaceae *152*
2. Coriander (Coriandrum sativum) – Umbelliferae *152*
J. Weed Control in Fodder Legumes 153
1. Lucerne (Alfalfa): (Medicago sativa) – Papilionaceae *153*
2. Berseem (Trifolium alexandrinum) – Papilionaceae *153*
K. Weed Control in Plantation Crops 153
1. Coffee (Coffea arabica) – Rubiaceae *154*
2. Tea (Camellia sinensis) – Theaceae *154*
3. Rubber (Hevea brasiliensis) – Euphorbiaceae *154*
4. Coconut (Cocos nucifera) – Arecaceae *155*
L. Weed Control in Horticultural Plants 155
1. Citrus (Citrus spp.) – Rutaceae *155*
2. Banana (Musa sapientum) – Musaceae *155*
3. Pineapple (Ananas sativus) – Bromeliaceae *156*
4. Grape (Vitis vinifera) Vitaceae, Plums (Prunus Domestica) – Rosaceae, Apple (Pyrus malus) Rosaceae, Gooseberry (Physalis peruviana) Solanaceae; and Strawberry (Fragaria vasca) Rosaceae *156*
M. Weed Control in Ornamental Trees, Shrubs and Flowering Plants 157
1. Ornamental Trees *157*
2. Nursery Stock *157*
3. Corms, Bulbous Ornamentals *157*

4. Gladiolous – Iridaceae *157*
5. Chrysanthemum (Chrysanthemum morifolium) – Compositae *158*
6. Roses (Rosa damascene) – Rosaceae *158*
N. Use of Herbicides in Forestry 158
1. Herbicides for Forest Nurseries *158*
2. Herbicides for Established Forests *158*
O. Choice of Weed Control in Cropping System 159
P. Weed Control in Grass Lands 160
Q. Control of Weeds in Lawns and Paths 160
1. In Newly Sown Lawns and Established Lawns *160*
2. Moss Control in Lawns *161*
3. Weed Control on Paths and Hard Surfaces (which are never dug or disturbed) *161*
R. Weed Control in Non-Cropped Areas 162
1. Control of Herbaceous Vegetation in Non-Agricultural Land *162*
2. Herbicides for Non-Cropped Land *162*
3. Basal Dark Spraying *164*
4. Scrub Control *164*
5. Sterilants Used in Non-Cropped Land *165*
6. Soil Fumigants Used in Non-cropped Land *165*

8. HERBICIDES 166
Herbicides – Classification and Selectivity 167
A. Selective Herbicides 168
B. Non-selective Herbicides 168
Soil Application *168*
Foliar Application *170*
Aquatic Application of Weedicides *171*
Methods of Herbicide Application *172*
Herbicides – A Chemical Classification 173
Inorganic Herbicides 178
Arsenic Acid *178*
Borax (Trade Name: Decahydrate): 100% *178*
Borate: (Sodium Metaborate) *178*
Copper Sulphate *178*
Ferrous Sulphate *179*
Sodium Chloride and Sodium Nitrate *179*
Sodium Chlorate and Sodium Arsenite (WP 100%) *179*
Sulphuric Acid *179*
AMS (Ammonium Sulfamate): (95%) *179*

Herbicidal Petroleum Oils *179*
Organic Herbicidal Compounds 180
Aliphatics *180*
Dalapon – Sodium Dowpon: WP 85% *180*
TCA (Varitox) – Trichloro Acetic Acid, WP 90% *181*
Amides and Anilides 181
Alachlor (Lasso) EC 48%, Gr. 10% *182*
Butachlor (Machete) EC 50% ai; & 5% G *182*
Propachlor (Ramrod) WP 65% *183*
Propanil (Stam F-34, Rogue) EC 34% *183*
CDAA (Randox) Gr 20% *183*
Pronamide (Kerb): WP 50% *184*
Naptalam (Alanap) Ec. 35%, Gr. 10% *184*
Acrolein (Aqualin): EC 92% *185*
Anilines and Nitro-Phenols 185
Dinitramine (Cobex): EC 25% *185*
Nitralin (Planavin): WP 75% *185*
Trifluralin (Treflan) EC 48%, Gr. 5% *186*
Fluochlorlalin (Basalin) EC. 48% *186*
Dinoseb (Pre-emergence, Dow Weed Killer): EC 48% *187*
PCP (Durotox) *187*
Fluorodifen (Soyex, Preforan): EC 30% *188*
Nitrofen (Tok. E-25): EC. 25% *188*
Arsenicals 188
DSMA (Ansar 529): EC 35% *189*
MAA (Ansar, Weed-E-Rad): EC 35% *189*
MSMA (Ansar-529, Daconate): EC 35% *189*
Benzoics and Phenyl Acetic Acid 189
Chloramben (Amiben, Vegiben): EC 24%, Gr. 4% *189*
Dicamba (Banvel): EC 50% *190*
Fenac (Fenac) EC 40% *190*
Carbamates, Thio-Carbamates and Carbonilates 191
Diclormate (Sirmate): EC 80%; Gr. 5% *191*
Asulam (Asulox): Ec 40% *191*
Barban (Carbyne): EC 25% *192*
Chlorpropham (CIPC): EC 40% *192*
Propham (IPC): WP 50% *192*
Thio-Carbamate 193
Benthiocarb (Saturn, Bolero): EC 50%; Gr. 10% *193*
EPTC (Eptam): EC 75%; Gr. 10% *193*
Diallate (Avadex): EC 40% *194*
Tri-allate (Avadex BW): EC 40% *194*

CDEC (Vegadex): Gr. 20% *194*
Molinate (Ordram-72): EC 72%; Gr. 75% *195*
Bentazon (Basagran): WP 48% *195*
SWEP *195*
Glyphosate (Round up): EC 41% *195*
Heterocyclic Compounds 196
Bipyridylium *196*
Pyridines *197*
Pyridazines *198*
Uracils *198*
Triazines 200
Atrazine (Atrataf, Residox): WP 50% *200*
Simazine (Gesatop, Prinlep): WP 50%, Gr. 20% *201*
Propazine (Gesamil, Milogard): WP 50% *202*
Ametryne (Gesapex): WP 80% *202*
Prometone (Pramitol): EC 25% *202*
Promatryne (Gesagerd): WP 50% *203*
Terbutryn (Igran): WP 80% *203*
Metribuzin (Sencor): WP 70% *203*
Misc. Heterocyclic Compounds 204
Aminotriazole *204*
Hormone-Type herbicides 205
a. Phenony Acetic Acid: WP 80% *205*
b. ∝-Phenoxy Propionic Acid *207*
γ-Phenoxy Butryric Acid *208*
Nitriles 209
Bromoxynil (Buctril): EC 35% *210*
Dichlobenil (Casoron): WP 45% *210*
Ioxynil (Actril): EC 25% *210*
Substituted Ureas 211
Chloroxuron (Tenoran): WP 50% *211*
Diuron (Karmex): WP 80% *211*
Fenuron (Urab): WP 50% *212*
Fluometuron (Cotoran): WP 50% *212*
Monuron (Telvar): WP 50% *213*
Alkoxy 213
Linuron (Lorox, Afalon): WP 50% *213*
Chlorbromuron (Maloran): WP 50% *214*
Neburon (Bonus): WP 60% *214*
Unclassified 214
Methazole (Probe): WP 75% *214*
Perfluidon (Destun) *214*

C-288: Gr. 2.5% *214*

9. WEED CONTROL TECHNOLOGY AND TRAINING **216**
Technology 216
Training 217

10. APPENDIX **219**
1. Relative Mobility Classes of Herbicides in Soils in Ascending Order from Left to Right (After Helting and Turner 1968) 219
2. Susceptibility of Crops on Certain Herbicides (After Hilton et al., WSSA 1974) 220
3. Organic Herbicides 221
4. List of Herbicides, Rate and Mode of Treatment 224
5. Herbicide Combinations 228
6. Selective Bibliography 230
7. Address of Chemical Companies and Source of Herbicides, Application Equipments and Weeding Implements 233
8. World's Worse Weeds 237
9. Terms and Abbreviations 239

CHAPTER 1

Definition and some Basic Concepts

1. Introduction

Weed is a plant growing out of place. They are unwanted non-useful, persistent and prolific, effectively competing with the beneficial and desirable crop plants for space, nutrients, sunlight and water; interfere with agricultural operations and thereby reduce the yield and quality of the produce. Certain species with pernicious or even poisonous characteristics, grow where they are not wanted.

1.1. Basic Concept

Weeds have existed from the beginning of agriculture. From the earliest periods of their existence, the primitive farmer had tried to pull them out by hand and this has prevented them from competing with cereal crops, even though this was a tedious agricultural operation. This simple means of eradicating weeds has been in vogue and is being carried out often by women and children. With the advancement of agricultural technology, mechanical weeding was found to be a quick and practical method of fighting weeds and now the chemical revolution is showing new heights of efficiency. With the adoption of this modern technology, weeds will no longer be a limiting factor in crop production. As labour becomes scarce and the wages are constantly increasing, the cost of weed control employing manual labour is getting uneconomical, and thus, chemical weed control is progressing at an accelerating rate under such situations which is helping the weedicide trade to come to the forefront, in recent years.

With the rapid improvement of crop production, certain fundamentals like weed control have been relegated to the bottom and not much time has been devoted to weed control technology, training and socio-economic aspects to solve the weed problem.

Significant steps towards a weed free agriculture have to be evolved so that the farmer will have more time for other gainful employment, by protecting the crops from serious damage by weeds. The approach to weed problem has to be considered in terms of the whole system which may pave the way towards a weed-free crop by adopting certain effective mechanical or chemical means of control.

The aggressiveness of weeds over crops inspite of man's best attempts to control them, can be attributed to certain powerful characteristics of weeds that has made them dominant over the crops.

1.2. Losses Caused by Weeds

Most of the weeds complete their life-cycle within a very short time when compared to the crops in which they occur. Losses caused by weeds exceed the losses from any other agricultural pests. It is estimated that among the annual agricultural loss in India, weeds account for 45%, insects 30%, disease 20% and other pests 5%. Thus, of the four groups of agricultural pests, the greatest losses are caused by weeds. According to Mani *et al* (1968), weeds incur a loss of 31.5% in food grain crops of which maximum loss in the kharif season. Reduction in yield due to weeds is maximum in sugarbeet (70.3%), followed by onion (68%), sugarcane (34.2%), linseed (34.2%), groundnut (33.8%), peas (32.9%) etc.

2. Effect of Weeds and their Competition

Weed competition is very much complicated because of various factors involved. Competition between crops and weeds is most severe when the competing plants are having similar vegetative habits and almost similar demand upon available resources. Hence, if weeds are not smothered at early stages, they become serious competitions in later stages and cause considerable reduction in crop yield. Weed competition mostly depends upon certain factors like, type of weed species and its duration, competing ability of crop plant, severity of infestation and especially soil moisture and climatic condition for its favourable growth.

2.1. Competition of Weeds for Water, Light and Mineral Nutrients

Weeds compete successfully with crops for water, light, space, and nutrients which affect growing of crop plants.

i. Competition for Water

Weeds generally absorbs and transpire more water than most crop plants such as jowar, bajra, maize etc. Certain weeds like ragweed (*Parthenium hysterophorus*), and common ragweed (*Ambrosia* spp.) require water to the extent of about three times that of millets. It is estimated that the requirement of jowar plant is about 300 litres of water; while the associated weeds like *Tephrosia purpurea* amounts to 1108 litre water and for *Tridax procumbens*, 1400 1 water. In fallow fields, weeds cause severe soil moisture depletion and transpire the available moisture rapidly.

ii. Competition for Incident Solar Energy

Light is an important factor for rapid growth of crop plants as well as weeds. Photosynthesis is dependent upon light. Broad-leaved weeds establish earlier to the crop plants and restrict the latter's photosynthetic activity through shading from the very beginning, thereby hindering crop growth. This dominance of weed association over crops in reducing available light is most pronounced in slow-germinating crops like groundnut, sugarcane etc. It is estimated that weed competition reduces light intensity by as much as 85 per cent in onions and beets, thereby reducing the yield by 60 per cent.

iii. Competition for Nutrients

Weeds remove from soil mineral nutrients like nitrogen, phosphorus and potash more efficiently than the crop plants and thus depress nitrogen and particularly potassium content of crops. Certain weeds have very deep and also prolific root system which check the normal nutrient absorption of certain crops and thus cause reduction in crop yield. Mani *et al.*, (1975) estimated that the weeds come up during Kharif season in crop fields, remove more N.P.K. (45 Kg N, 12 Kg P_2O_5, 73 Kg K_2O/ha) than the rabi season weeds found in wheat field which remove 24 kg N, 4 kg P_2O_5 and 26 kg K_2O/ha. Some parasitic weeds like dodder (*Cuscuta* sp.), *Loranthus* sp. etc. absorb mineral nutrients directly from the host crops and destroy them in the long run. Certain weeds accumulate high quantity of potash and nitrogen also. *Amaranthus viridis*, *Chenopodium album*, *Portulaca quadrifida*, *Celosia argentea* can remove 3.5 to 4.5% K_2O and 2.4 to 3.1% N from the soil during the crop period and thus become a great competitor for nutrients.

2.2. Competition of Weeds for Space

Weeds compete for space both in the rhizosphere and atmosphere. They restrict the root growth and volume of the cultivated crop plants. As a result, crop plants absorb less moisture and mineral nutrients from the soil in the weed-infested areas, resulting in heavy loss of crop yield. Further, due to heavy competition from weed associations, crop plants get only limited space to develop their shoot system which affects their photosynthetic activity adversely.

2.3. Weeds Increase Crop Pests

Weeds host many pathogens and insect pests in off-season which migrate to the crop later and cause severe damage. Grasshoppers and nematodes live and multiply on weeds and thereby cause damage to many crop plants. Sweet potato weevil (*Cylas formicarius*) infests field bindweed (*Convolvulus arvensis*) and later becomes a major pest during sweet potato season. Weeds like *Solanum* spp. and *Chenopodium album* are the common hosts for stalk borer, beetles and cutworm which later migrate to crops like potato, tomatoes, maize, gram, peas etc. and damage them severely. *Physalis minima* hosts mosaic disease, cyst nematode, stalk borer and beetles in off-season; which later migrate to tomatoes and potatoes.

2.4. Weeds Interfere with Crop Culture

In weedy fields, application of fertilizers or providing supplementary irrigation become very cumbersome. Certain twining weeds, like bindweed (*Convolvulus arvensis*), get entangled with crop plants very badly, thus creating difficulties for harvest of the mature crops, besides restricting the growth of the host crop.

2.5. Weeds Reduce Crop Quality

Parasitic weeds like *Striga lutea* reduces the quality of sugarcane juice; weeds like nutsedge (*Cyperus* spp.), make the hay or straw less palatable to animals; similarly, wild onion or wild garlic (*Alliums* spp.) mixed in forage crops, impart off-flavours to milk. Weed seeds like wild mustard (*Brassica nigra*), Mexican poppy (*Argemone mexicana*) mixed with wheat grains or edible mustard cause objectionable odour to the flour and can even prove poisonous. The weed, cocklebur (*Xanthium strumarium*) seriously impairs the quality of sheep wool.

2.6. Weeds Harm Animal Health

Several weeds prove poisonous to animals when ingested, because they contain toxic alkaloids, cyanogenic glucosides, oxalates, nitrates etc. Weeds like poison ivy (*Rhus radicans*), poison oak cause severe itchy rashes and dermatitis; many others cause hay fever and allergic reactions. In some natural and neglected grassland, many poisonous weed species grow and cause harm to grazing animals.

Lantana sp. induces hyper-sensitivity to light in animals feeding upon it. Johnson grass (*Sorghum halepense*) at its tillering phase, cocklebur (*Xanthium strumarium*) at its emergence and congress weed (*Parthenium hysterophorus*) throughout its life period, are very poisonous to animals and man. Sweet clover (*Melilotus alba*) when fed in larger quantity acts as anti-blood coagulant.

2.7. Weeds Harm Human Health

Many weeds are responsible for human health problems and cause allergic reaction. Poisonous weeds like congress weed (*Parthenium hysterophorus*) poison oak, poison ivy, cause allergy on direct contact, severe itching and inflammation.

2.8. Allelopathic Effects of Weeds

The phenomenon of one plant having detrimental effect on another through the production of certain chemical compounds is called "allelopathy". The allelopathic effect depends upon excretion of toxic substances from their roots which affect their neighbours. The liberation of such endogenous substances like kaolins, lactones amino acids and peptides by plant roots, cause inhibitory influences on cultivated plants. The weeds like quack grass (*Agropyron repens*) is able to inhibit growth of crop seedlings; wild oat (*Avena fatua*) has inhibitory influence on other plants; nut sedge (*Cyperus* sp.) causes stunting in growth of cotton.

2.9. Effect of Parasitic Weeds

Some of the most serious weeds are parasitic upon crops. Among parasitic weeds, broom rape (*Orobanche aegyptica*) and figwort (*Striga lutea*) are angiosphermous root parasites that grow among the tissues of the host plants. They grow in the presence of substances that escape from the roots of the host plants like tobacco (in case of *Orobanche*) and bajra or jowar (in case of *Striga*). Other parasitic weeds are *Loranthus* sp. Which is mostly found in mango orchards and also other trees, and dodder (*Cuscuta* spp.) which appear in bushes and neglected gardens. *Loranthus* is causing serious damage particularly to mango trees in recent years.

2.10. Weeds Contaminate Water Body

Aquatic weeds change the taste of drinking water. Free-floating weeds like water-hyacinth, *Salvinia*, *Pistia* form large mats and hinder navigation, choke irrigation channels and drainage, interfere with swimming, boating, fishing and hampers growth of wetland rice and when they decompose partially, they contaminate water-body very hardly.

2.11. Weeds Interfere in Non-Cropped Lands

Weeds spread wildly on rail tracks, road sides, cling to fenes, pipelines, poles and covers drainage channels; shrubs grow profusely on waste land and forest areas. Thus, weeds become a great menace to non-cropped lands and make the area messy.

3. Classification of Weeds

3.1. According to Ontogeny of Weeds

There are three principal groups of weeds according to their life-cycle viz. annuals, biennials and perennials. Weeds belonging to any of these classes have specific mode of propagation, dispersal and persistence. In each group, there are broad leaved and grassy weeds.

i. Annual Weeds

Annual weeds are most common in the fields, which complete their life cycle in one season or in less than a year. They are very persistent because of their abundant seed production and fast growing habit and hence it becomes very costly to control them. Annuals are of three types: (i) Summer (ii) Kharif and (iii) Winter annuals.

Summer and Kharif Annuals

The summer and kharif annuals are usually the same plant species which are most troublesome to seasonal crops like sorghum, groundnut, cotton, tobacco and many vegetables. These weeds include: pigweed (*Amaranthus* spp.), ragweed (*Ambrosia* sp.), *Digera arvensis*, Purslane (*Portulaca oleracea*), crab grass (*Digitaria* spp.) goose grass (*Eleusine* spp.), foxtail (*Andropogon* sp.) etc.

Winter Annuals

Winter annuals germinate in winter and mature seeds in early summer. Seeds lie dormant in the soil during summer months due to high soil temperature. Winter annuals include lambs quarters (*Chenopodium* sp.), vetches (*Vicia* sp.) etc.

ii. Biennial Weeds

Biennial weeds like wild carrot (*Daucus carota*), alligator weed (*Alternanthera philoxeroides*) complete their life-cycle in two years; remaining vegetative in the first year and flowering and setting seeds in the second year.

iii. Perennial Weeds

Perennial weeds live more than three years and reproduce by seeds or spread vegetatively. They start flowering from second year onwards. Some perennial weeds propagate through under-ground parts like bulbs, tubers etc. like nutsedge (*Cyperus* spp.) some are reproduced by above ground stolons like Bermuda grass (*Cynodon dactylon*), quack grass (*Agropyron repens*) which are shallow rooted; while deep-rooted ones like Johnson grass (*Sorghum halepense*) reproduce both vegetatively and by seeds.

3.2. According to Cotyledon Characters

Weeds are broadly divided into three: (i) Grasses (ii) dicots and (iii) sedges. The grassy weeds are narrow leaved, dicots are broadleaved; while sedges (*Cyperus* sp.) and cattails (*Typha* spp) etc. through narrow leaved, are not grasses.

3.3. According to Nature of Stem

Weeds can be classified into three: (i) herbaceous (ii) semi-woody and (iii) woody species. Herbaceous weeds have green and succulent stems. Semi-woody weeds are under shrubs like *Lantana camara* and woody weeds include acacia (*Acacia* sp.) etc. Both semi-woody and woody species are collectively called ‘brush’.

3.4. According to Soil

On highly acid soils, wood-sorrel (*Oxalis* sp.) can dominate; while on saline or alkali soils, salt grass, alkali grass, smart weed (*Polygonum* sp.) can flourish well.

3.5. According to Climate

Weeds grown in hilly areas are quite different from those in the plains. In temperate climate, annual weeds tend to be prominent; while in the humid topics, perennials dominate.

3.6. According to Association

Weeds mostly grow in three types of associations: (i) Season bound (ii) Crop bound and (iii) Crop associated. Season bound weeds grow in specific season of the year like *Chenopodium album* grown in winter (ii) crop bound are parasitic and depend for their survival upon their host plants for certain stimuli and nutrition (like *Striga* on bajra or Orobanche on tobacco) (iii) Crop-associated weeds are crop specific such as wild rice in rice field, wild oat in small grains.

3.7. According to Situation

Occurrence of weeds depend upon nine types of situations crop land, fallow land, grass land and pasture non-crop land, aquatic weeds, forest land, lawn and gardens, orchards and plantations.

4. Ecology of Weeds

4.1. Weed Ecology

Weed ecology is the inter-relationship between the growth characters and adaptations of weeds to the changes in their environment. The environment includes climatic, adaphic (soil) and biotic factors. These factors affect the occurrence, abundance distribution and behaviour of weeds. Weeds can grow under varying soil and climatic conditions and also survive drought, frost and high temperature. They appear almost everywhere, in crop lands or non-cropped areas.

i. Climatic Factors

Distribution of weeds is closely related to climate, including light, temperature, rainfall and humidity. Some weeds thrive under any conditions while others are highly selective in their growing conditions. Weeds like nutsedge is most commonly grown throughout humid tropics. Some weeds like pigweeds (*Amaranthus* spp.) prefers mesophytic conditions; while prickly pear (*Opuntia* spp.).

ii. Soil Factors

Soil factors which influence the persistence of weeds are soil moisture, aeration, soil temperature soil pH and soil fertility. Cropping systems also play an important role. Very few weed species are associated with any particular type of soil. However, some of the weed species act as indicator plants according to reactions of the soil; like alkali- tolerant weeds might indicate the suitability of growing sugarbeet successfully and its unsuitability to alkali-susceptible maize plants. In certain depleted and eroded soils, quackgrass (*Agropyron* sp.) Canada thistle (*Cirsium arvense*), wild carrot (*Daucus carota*) and found abundant.

iii. Biotic Factors

Plants and animals are the biotic factors that affect weed persistence. Crops that serve as hosts like tobacco to parasitic weeds like Orobanche and crop induced toxins like Kaolines exuded from roots are some of the biotic factors.

4.2. Survival Mechanism of Weeds

Annual weeds are principally reproduced by seeds; while the perennials survive both vegetatively and as seeds. Weeds produce enormous number of seeds which vary widely among species. Wild oat (*Avena fatua*) can produce about 250 weeds per plant, while a pigweed (*Amaranthus* sp.) and purslane (*Portulaca oleracea*) can bear millions of small seeds. The vegetative plant propagules such as bulbs, tubers, rhizomes, stolons etc. offer on additional survival mechanism for perennial weeds.

Weed seeds multiply profusely and spread rapidly with the help of certain morphological appendages and also through various agencies like human, cattle machineries, wind, water etc. Dormancy of weed seeds is one of the most important characteristic that enable them to survive for a long period, even number of years. The prolonged dormancy is due to a hard seed coat, improper development of embryo, barrier to prevent oxygen entry etc. hence, under unfavourable conditions, they lie in the soil for a long period; but under favourable moist soil conditions, they normally germinate before the germination of crop seeds and thereby smother the young crop seedlings.

5. Dispersal of Weeds

Weed seeds and fruits are dispersed in various ways. Weed seeds are mostly disseminated through: (i) a successful agent and/or (ii) by special adaptation. Weeds are most widely spread through crop-seeds and farm products. Wind, water, animals and farm machinery are the chief agencies for dissemination of weed seeds and hence, it is very difficult to control these natural agencies.

5.1. Wind as an Agent

Weed seeds have special adaptations and possess specific organs that keep them afloat through wind. Those seeds have structural modifications like wings, parachute, chaffy etc. which aid in their wind distribution to a longer distance. Winged fruits are seeds of umbelliferae; parachute fruits with tuft of hairs in compositae like Canada thistle (*Cirsium arvense*); comate seeds with silky hairs like milk-weed (*Calotropis* sp.) and prickly poppy (*Argemone mexicana*); chaffy seeds of Solanaceae like groundcherry (*Physalis minima*) etc. disperse their seeds some distance away by special adaptations.

5.2. Water as an Agent

Some weed seeds have species structure to help them float is water and thus are carried by water. Irrigation water is an important means of their rapid dispersal. Some seeds float on natural streams for days together and remain viable. Those weeds flavouring dispersal by water include: pigweed (*Amaranthus* spp.) lamb's quarters (*Chenopodium album*), bindweed (*Convolvulus* sp.) etc. Most aquatic weeds disperse largely through water currents or by force of downward stream.

5.3. Animal and Man as Agents

Animals carry the weed seeds in their feet, fur or by carrying ingested seed through digestive tracts which remain viable. Fleshy fruits like poison-ivy and poison-oak, are often eaten by birds or animals and carried way. Lantana seeds are dispersed through Indian mayama. Puncture vine (*Tribulus terrestris*), Cocklebar (*Xanthium strumarium*) etc. are dispersed through animal hooves, hairs etc. Commercial seeds contaminated with weed seeds are dispersed from one place to the other easily by man.

5.4. Farm Machineries Carry Weed Seeds

Weed seeds are carried on wheels of carts, farm machines and drop them elsewhere where new infestation starts. Dissemination of wild oat (*Avena fatua*) canary grass (*Phalaris minor*), wild rice (*Oryza* sp.) etc. which are largely seed contaminants are, dispersed easily through different media.

5.5. Manures Carry Weed Seeds

Farm yard manure or compost is the important source of dissemination of weed seeds. Field bind weed (Convolvulus sp.) seeds remain viable in compost for a long time. While manuring, FYM or compost are spread all over the field and incorporated along with weed seeds mixed in it which creates a great problem later for eradication.

6. Prevention, Eradication and Control of Weeds

6.1. Prevention of Weeds

Preventive measures are the practical means of controlling weeds, making sure that weeds seeds are not carried from one place to another and also preventing the weed spread on the farm through seed or reproduce vegetatively. As legends says 'an ounce of prevention is worth a pound of cure'. Introduced perennial weeds like water hyacinth have become most serious pests which are difficult to eradicate. Hence, efforts are needed to prevent the introduction of a weed or to prevent its spread, if sparsely appears.

Most common preventive methods are: Use of clean seeds, through cleaning of agricultural implements before moving them from infested areas; not to feed grain or hay containing weed seeds to the animals; to keep the banks of irrigation channels free from obnoxious weeds etc. Seed laws may prevent seed contamination.

6.2. Eradication of Weeds

Strict vigilance and well-planned long term programme have to be designed to eradicate the existing stand of perennial weeds. Eradication means complete elimination of both living weed seeds and the seeds present in the soil. Soils sterilants may be used for complete eradication in non-cropped and bare lands.

Eradication of noxious weeds like *Lantana*, *Eupatorium*, *Parthenium*, *Cuscuta*, *Striga* etc. is possible when the infestation is of limited areas; but when such weeds invade large areas, it becomes uneconomical for their eradication. Hence at their early stages and also while spread only in a limited area, those should be eradicated through voluntary squads or herbicidal control means.

6.3. Weed Control Methods

Weed control is the process to limit the growth of unwanted plants mostly from cultivated fields. Mechanical methods must be integrated with appropriate herbicides in overall operations to make it more effective and cheap.

Methods of weed control may be classified into four groups:

1. Physical or mechanical methods, like hand weeding, hoeing, tillage, mowing, flooding, burning, smothering etc.
2. Rotational cropping and crop competition methods.
3. Chemical methods; using selective or non-selective herbicides contact or translocated, foliage or soil incorporation; trunk injection, application in water for aquatic weeds.
4. Biological methods.

6.3.1. Physical Method of Weeds Control

Physical methods include both manual and mechanical methods including hand-weeding to several means of mechanical control of different weeds. Though some of the method are primitive, those are most practical means and very safe to the crops. Hand tools or animal-power operated weeders are well known to the farmers and hence special technical skills are not needed. However, hand-weeding required high labour input; labour also becomes scarce for such weeding operations during peak periods when sowing, transplanting, harvesting of other crops coincide. Timely hand-weeding or using hand-hoeing tools in row crops are most practical and efficient methods to eliminate scattered weeds particularly in millets, cotton and pulse crops, though these common practices are labour intensive and time consuming.

Tillage operations can eliminate annual, biennial and perennial weeds; mowing is done to prevent seeding of all kind of weeds in waste lands. Flooding is popular for controlling noxious perennials grown it uplands; while smothering is done with mulching materials.

6.3.2. Rotational Cropping and Crop Competition Method

Rotation of different crops breaks the cycle of weeds and intensive cropping reduce the weed pressure.

Tall crops which have fast canopy forming ability, suffer less from weed competition than slow growing short stature crops like groundnut. Faulty germination and wide gap cause weed crop competition intense. Weed seeds germinate readily while crop mergence at longer intervals, which leads to severe weeds crop competition. Hence adequate seed rate, use of good quality seed and growing of quick germinating crop in weed susceptible areas, have to be practiced. Competitive cropping and rotational cropping reduce the germination and growth of weeds, specific to particular crop like, *Striga* on bajra or sorghum.

Sorghum, black gram etc. are the principle competitive crops. Maize + cowpea or maize + green gram shows 40% reduction in weed weight, while sole maize crop may not yield at all due to weed competition.

6.3.3. Biological Method of Weed Control

Biological control of plants by insects or fungus that live on specific weeds, is natural process which is harmless to desired plants. However, complete eradication is not possible by this method and hence an equilibrium for suppression of weed spread is needed. Biological methods usually control rather than eradicate.

To control certain alien weeds, Biologists employed insects which is called biological method of weed control. The control of Prickly pear (*Opuntia* spp.) by the moth borer (*Cactoblastis cactorum*) in Australia and that of thorny shrub, lantana (*Lantana camara*) in Hawaii with certain insect bioagents like cochineal scab insects, *Dactylopius tomentosus* and larvae of tortricid moth (*Crocidosema lantana*) which bore into stem, eat flowers and fruits, are the spectacular examples. More recently, alligator weed (*Alternanthera philoxeroides*) has been brought alligator weed (*Alternanthera phylloxeroides*) has been brought under biological control with flea beetle (*Agasicles hygrophila*) larvae which feed on leaves and finally bore into its stems to pupate inside.

Besides insects, certain fish carps, like common carp (*Cyprinus carpio* L.) and Chinese grass carp are promising species for aquatic weed control. Snails, mites and fungi have also been employed for the control of some aquatic and terrestrial weeds.

6.3.4. Chemical Methods of Weed Control

Even in developing countries like India, chemical weed control has not yet received much attention for various reasons. The farmers still rely on traditional methods of hand weeding and the agricultural labour is also available at a reasonable rate in the rural areas of many states. Further, the farmers are not properly trained about herbicides and their operation; in addition the cost of certain weedicides and their application are rather high. Hence, chemical weed control is still in its infancy, compared to the development in other agricultural sciences, like use of pesticides and fungicides.

However, in recent years, farmers have realized that they cannot afford to lose time on the time-consuming manual weed control when intensive and multiple cropping programme is followed and hence desired to control weeds in the early stage of crop growth by applying simple herbicides, particularly in row-crop production.

The weeding efficiency has thus greatly improved by supplementing conventional weeding methods with herbicidal applications either pre-emergence or post-emergence. The herbicides developed are in common use for selective and non-selective weed control in different areas. Use of selective pre-emergence herbicides for sole crops or mixed-population, can injure specific weeds and thereby help to increase the total hectare handled by a single family. However, careful evaluation is needed to see their residual effect to high intensive cropping system.

7. Formulation of Weedicides and Calculations

7.1. Formulations

Herbicides are formulated and applied as: (i) Solutions of water or oil, (ii) Emulsions, (iii) Wettable powders and (iv) Granular materials.

1. **Solutions:** A solution is a homogenous mixture of two or more substances which cannot be separated by physical means like salts in the solute. Most of such salts are soluble in water.
2. **Emulsions:** An emulsion is one liquid dispersed in another liquid, each maintains its original identity and hence the liquid may separate. Emulsion appears milky and constitute the oil-soluble herbicides like CDAA.
3. **Wettable powder:** A wettable powder suspension consists of finely divided solid particles dispersed in a liquid. Most suspensions require agitation to prevent setting of the solid particles. Wettable powder include atrazine, simazine, monuron, diuron etc.
4. **Granules:** Granular herbicides are spread by hand or

mechanical spreaders. It is generally applied by mixing with a carrier like sand, to make it bulk for even distribution. Granules are easy to apply without water or using spray equipments. Machete, bendiocarb etc. are available in granular form.

Spray drift and volatility: Herbicides may drift through the air which depends upon size of droplets, movement of wind and releasing spray from certain height.

Volatility refers to vaporization of chemicals into fumes. These vapours can damage susceptible crops.

7.2. Calculations

i. Active Ingredient

Active ingredient (a.i) is that constituent of chemical formulation which is directly responsible for the herbicidal effects. It is normally displayed by the manufacturer on herbicide formulation containers. It may be given as per cent by weight or volume (W/V). If a herbicide concentrate marked 80% W/V and it is decided to apply 1 kg ai/ha; one must use $\frac{100 \times 1}{80}$ or 1.25 kg of the material.

ii. Acid Equivalent

Acid equivalent (a.e.) refers to that part of formulation that theoretically can be converted to the acid. The herbicide manufacturer displays active ingredient (ai) of acid type of herbicide concentrates in terms of acid equivalent (a.e.) in the container like weder-96; Tok-E 25 in herbicide activity is more accurately calculated on an acid equivalent (a.e.) basis which is marked on the labels.

iii. Parts Per Million

Parts per million (ppm) refers to the number of parts by weight or volume of a constituent in 10,00,000 parts of the final mixture, by weight or volume.

iv. Percentage Concentration

Per cent concentration is similar to ppm except that it is expressed as a percentage.

$$\text{Per cent ppm} = \frac{\text{ppm}}{10{,}000} \quad 1 \text{ mg/1} = 1 \text{ ppm}$$

$$1g / 100\ ml = 1\%$$
$$\text{or ppm} = \text{ppm per cent} \times 10{,}000$$

8. Weedicides Application Equipments

Weedicides are mostly applied as solution by spraying, however, recently granular applications are gaining importance, particularly to apply in moist soil or low-lying rice fields. Careful spraying is utmost essential for successful results, because thorough absorption of herbicide by plants is a must.

8.1. Spraying Method and Spray Volume

Spraying of herbicides is a common method which is done uniformly and accurately to a given area, either in the soil or as foliar spray on the plant.

Normally two kinds of sprayers are used based on size of the droplets sprayed in low volume and high volume sprayers. Low volume sprayers apply upto 350 1/ha; while high volume sprayers apply 500 to 5000 1/ha. In case of translocated hormonal weedicides like 2,4-D, MCPA etc. low volume sprayers can be used. In case of contact herbicide like paraquat, high volume sprayer should be used for thorough 'wetting' and complete coverage of the plant foliage. In both cases, uniform application is required. The correct nozzle size and pressure should be chosen for the purpose.

Spray volume is the total quantity of liquid size applied per unit area of the target weeds or soil. As per USDA system, Ultra Low Volume (ULV) = upto 5.6 1/ha; Low volume (LV) = 12 to 125 1/ha, Semi-low volume = 125 to 500 1/ha and High volume = 500 or more L/ha.

8.2. Parts of the Sprayer and their Functions

Any sprayer consists of three essential parts: (i) tank for holding spray fluid; (ii) pumping mechanism and (iii) spraying apparatus consists of plastic tube, lance and nozzle with governor mechanism. Other essential accessories are: filter or strainer, pressure gauge, pressure regulator, shut off valve and connecting hose.

i. Tank

Tank is usually made of brass or polythene. In case of pneumatic sprayer, high pressure is not allowed to develop. It holds the spray fluid.

ii. Nozzle

The nozzle is the most important part of the sprayer. According to design of nozzle, spray drift, uniformity of application and rate of application are regulated. The nozzle converts the spray liquid into spray droplets. Compared to brass nozzles, plastic nozzles are more resistant and cheaper. At low pressure, the liquid escapes from the nozzle tip as a liquid form; while under high pressure, droplets turn to fog or mist size with the formation of very fine droplets. Full cone type nozzle is particularly useful for spot spraying on a smaller area. Since the nozzle is the most important part of the sprayer, this should be maintained properly by thorough cleaning before the after spraying and should be replaced once in a year.

iii. Pump

Pumps are of two types – One delivers gas or air pressure and the other liquid pressure. The spray tank should be strong enough to withstand the pressure. Since the gas or air pressure system costs more, liquid pressure system is normally used. In liquid pressure system, pressure passes form the pump to nozzle through the boom. Five common types of liquid pressure pumps are used *viz*; centrifugal or turbine, gear pump, impeller, diaphragm and piston pumps.

1. Centrifugal or turbine pumps develop pressure as a result of centrifugal force; rotate rapidly and produce liquid velocity which gives its pressure. The liquid is released at the outer edge of the pump.
2. Gear pumps depend upon the misting of gear teeth to develop both vacuum and pressure.
3. Impeller pumps consist of a motor set to one side within the pump housing.
4. Diaphragm pump can handle sprays which are both chemically and abrasively corrosion.
5. Piston-type pumps can produce very high pressures. They are reliable and long-life pumps.

8.3. Cleaning the Sprayer

Cleaning the sprayer and rinsing the tank just after use is essential. In case of water-soluble herbicides, rinsing with water is enough, but for oil soluble herbicides like 2-4-D (ester), a wetting agent or detergent has to be added in the water for rinsing. Activated charcoal 0.3% suspension can also be used after preliminary rinsing.

This washing water should be spread on waste land but never to the fish pond or cultivated field.

The blocked nozzle should not be cleansed by sucking or blowing with the mouth. Protective clothing and face mask have to be used while mixing the chemical or spraying. Thorough washing is required afterwards.

9. Safety Precautions for Handing Herbicides and their Applications

Users should have considerable knowledge about herbicides and should take enough precautions while transporting herbicides from one place to another; also during its storage along with other materials and at the time of mixing for application in the field. Different herbicides cause several symptoms of poisoning like skin and eye irritation, inflammation, nose bleeding, weakness and lethargy, cardiac arrest etc. Swallowing results ulceration of mouth and throat, looseness of bowels, abdominal discomforts etc.

1. Herbicides should be transported in tightly close containers and enough precautions is to be taken in case of volatile herbicides. Carrying such chemicals on shoulders should be avoided.
2. Those chemicals should be kept away from food or feed stuff in the store. Its contamination with seeds, fertilizers, fungicides and insecticides should be avoided. After its use partly, the containers should be kept at air tight to avoid inactivation. Unlabeled containers should be discarded. Periodical checking in the store is required from any leaks from containers.
3. Containers should be kept out of touch of children.
4. Inhalation of spray mist, contact with skin or eyes or clothing should be avoided.
5. To blow out clogged-nozzle of sprayers, mouth should never be used by mistake and swallowing should be avoided. Mist blower should not be used for spraying in crop areas. Any drift to nearby desirable plant should be avoided.
6. During mixing of herbicides, rubber gloves should be used.
7. Empty containers and wastes should be promptly disposed off by burying in a waste land or a safe place.
8. Water bodies should not be contaminated by discarding left-over chemicals or washing sprayers on irrigation channels or

tanks or run-off is likely to occur.

9. Glass wares and spray equipments should be flushed thoroughly with water in a safe place, after use of chemicals.

First-Aid and Antidotes

External: If spilled on the skin, remove all contaminated clothing, clean with soap and ample running water. Wash contaminated clothing before reuse.

If the materials get into the eyes, wash immediately with running water for at least 15 minutes. After flushes, irrigate for a prolonged period. If irritation develops, get immediate medical attention.

Internal: If the material is swallowed, induce vomiting immediately by giving a spoonful of salt in a glass of warm water. This may also be done by introducing a finger into the throat. Administer cold milk or cold water freely and again induce vomiting until fluid is clear.

Treatment for ingestion: Gastric lavage, saline cathartic such s sodium sulphate, forced diuresis and symptomatic treatment, if breathing is stopped, quindine sulphate to counter mycotonia would be the immediate precautionary measures.

10. Absorption and Translocation of Herbicides

10.1. Physiological Response to Herbicides

Physiological processes of plants is modified even with small amount of growth regulators or hormones which either promote or inhibit the growth of plants depending upon the quantities used. Growth regulator like 2,4-D are translocated through the entire plant, effecting the physiological systems. Physiological differences in plants in the enzyme systems, cell metabolism etc. may result in blocking or stimulating certain biochemical process.

10.2. Morphological Responses to Herbicides

Herbicide application disturbances in cell division and tissue morphology. Accelerated pro-cambium activity in leaves resulted in destruction in phloem parenchyma, sieve elements and companion cells. All chloroplasts in treated plant leaves become devoid of starch; Photosynthesis is evidently inhibited.

The effects of translocated herbicides like 2,4-D in high concentration are the cell proliferation, starting in the endodermis of the stem and spreading to phloem and cortex parenchyma until the phloem is completely crushed. Contact herbicides like paraquat ruptures chloroplast membranes and prevents starch formation.

Morphological differences helps selection of herbicidal application, protect meristematic regions from herbicidal injury and affect retention of chemicals because of plant surface differences.

10.3. Absorption and Translocation

The outer cell wall of foliage is made up of four distinct substances viz. cutin, cuticle wax, pectin and cellulose. Cutin is strongly absorbent of ultra-violet light, which affords protection to photosynthetic mechanism of plant cells. Cutin was does not absorb ultra violet radiation. Pectins are amorphous and highly hydrophilic, capable of forming salts. They do not absorb ultra violet light. Cellulose is responsible for high resistance of cell walls of stretching.

Turgor results in enabling the plant to grow erect against the force of gravity. Stomata are openings in the cuticularised epidermis that allow gaseous exchange to go in so that carbon dioxide can diffuse into leaves.

10.3.1. Mechanics of Absorption

Penetration and absorption of the herbicides are the first process involved in herbicide physiology. There are two routes by which exogenous molecules may traverse the distance from cuticle surface into the living inner cells, a lipoid route and an aqueous path-way. Penetration of both types of compounds is greatly enhanced by providing a saturated atmosphere around the leaves. Differential absorption may account for differences in plant responses. Some plant surface absorbs the herbicides quickly while others absorb slowly.

When the plant is under stress for water, the water in the pores of the cuticle recede, the pores fill with air, and a droplet falling on the leaf would entrap air bubbles in the pores.

The two most common points of entry in plants are through the leaves or through the roots, mostly penetrated through the leaf surface. Spray solution absorbed through open stomata into the stomatal chambers, would be in a nearly saturated environment; probably the aqueous content would be much greater. This may be one of the principal advantages of stomatal uptake.

Thus the chemicals penetrates the layer of wax-like cuticle and cell walls which are chemically non-polar materials. Hence non-polar herbicides like 2,4-D (ester), Dinitraphenol tend to absorb in the leaves faster than the polar herbicides.

Roots are best adapted to absorb polar substances like 2,4-D. Rapid growth of roots favour rapid nutrient absorption and have rapid rate of respiration. A non-polar herbicide like ester of 2,4-D applied to the soil may be quickly converted to a polar substances like 2,4-D and further converted to a salt of 2,4-D.

10.3.2. Mechanics of Translocation

Translocation of herbicides is a major problem in those weeds which have below-ground reproductive organs. Translocation within the plant is done through the phloem or through the xylem or intercellular translocation.

1. Translocation through the phloem usually occurs in case of foliar spray of growth regulators like 2,4-D. Phloem tissues are composed of living cells and hence when the chemical is sprayed on the foliar, it is readily absorbed and moves from the leaves towards the roots upto 10 cm within 5 minutes. Normally, the chemical moves faster in young plants than the old ones. Translocation of 2,4-D becomes much faster in plans rich in potassium or phosphorus.
2. Translocation through the xylem is the reverse process in which the herbicides applied in the soil, is absorbed from the soil by roots and quickly translocated to all parts of the plant, without any injury to the xylem while moving.
3. In case of intercellular translocation, oil may be observed through the cuticle and the movement of confined to intercellular spaces.

11. Effect of Herbicides Applied on Soil

Herbicides are applied to the soil directly as: (1) Pre-planting treatment, (2) Pre-emergence (3) Post-emergence and (4) Soil sterilants.

As a pre-planting treatment, some chemicals are applied to the soil prior to planting, either mixed into the soil or applied to the surface.

Pre-emergence application requires a high concentration of an herbicide in the upper 1.5 cm of soil, where most annual weed seeds germinate.

In case of soil sterilization, the chemical must remain active in the rooting zone of germinating seeds, or deep-rooted plants.

11.1. Persistence in the Soil

Activity or persistence of herbicide in the soil relates to the lengths of time upto which the weed control is expected. Residual toxicity relates to phytotoxic after-affects, which causes injury to succeeding crops.

Six factors affect the persistence of an herbicide in the soil (a) micro-organism decomposition, (b) chemical decomposition, (c) photo-decomposition, (d) absorption on soil colloids, (e) leaching (f) volatility.

a. Macro-organism Decomposition

In the soil, microorganisms like algae, fungi, actinomycetes and bacteria can decompose some herbicide easily. A warm, moist, well-aerated and fertile soil is most favourable to micro-organism for quick decomposition of organic herbicides. In this process, oxygen is absorbed and carbon-di-oxide liberated.

b. Chemical Decomposition

Chemical decomposition is the process of oxidation, reduction, hydrolysis and hydration. Such process destroys herbicidal effects.

c. Photo-Decomposition

Photo-decomposition is observed in some herbicides like monuron which cannot remain for a longer period on the soil surface without rain or sufficient moisture on the surface soil.

d. Absorption on Soil Colloids

Those soils having high clay colloids and organic matter require large quantity of pre-emergence and soil-sterilant herbicides for weed control. Such sols also tend to hold the herbicides for a longer period than sand and hence, rate of herbicides use is correlated with absorptive capacity of various soils.

e. Leaching

Rain leaches the chemicals into the soil layers and hence pre-emergence herbicides are frequently applied to surface soil. The effectiveness of a herbicides depends on its movement by leaching and thereby killing of weed seeds germinating in the soil surface.

f. Volatility

Herbicides like ester-form of 2,4-D, EPTC are volatile which may be lost to the atmosphere as volatile gases.

The herbicides like EPTC may move into the porous soil as a volatile gas and effectively kill germinating seeds. Those herbicides once absorbed by the sol, usually reduce the loss of volatility.

CHAPTER 2

Monocot Weeds

Monocotyledons comprise the largest group of flowering plants in which almost all are herbaceous with hollow internodes and jointed nodes. The embryo is with one cotyledon, stem with vascular bundles usually closed and scattered, leaf venation usually parallel and flowers with trimerous perianth.

Grasses – Gramineae

1. *Agropyron repens*

Quack grass or common couch.............................. Gramineae

Quack grass is a shallow-rooted perennial weed. The rhizomes are spread in shallow depth of soil and the reproduction through rhizomes is a common feature. Its rhizomes are sometimes called 'soboles'. It can easily be killed at early stage, upto 6 weeks age; but after that seedling stage, it produces rhizomes.

It is having narrow leaf blade, leaf-sheathe usually with small auricles; spikelets many-flowered mostly bisexual, alternating on opposite sides of jointed rachis is solitary spikelets at each node; glumes well-developed.

Allelopathic affect: As a result of its decay in the soil, inhibitory action is produced by the root system of previous quackgrass which interferes the growth of maize, wheat and pea seedlings.

Monocotyledonous weeds (alphabetic order)

Scientific name	Common name	Family
Agropyron repens (L) Beaur	Quack grass	Gramineae
Agrostis palustris (Huds) Pers.	Bent grass	Agrostideae (Gramineae)
Allium canadense L.	Wild onion	Liliaceae
Allium vineale L.	Wild garlic	Liliaceae
Andropogon annulatus Orsk.	Fox-tail grass	Andropogoneae
Aristida dichotoma Michx	Threeawn	Agrostideae

Scientific name	Common name	Family
		(Gramineae)
Asphodelus fistulosus L.	Jungle onion	Liliaceae
Avena fatua L.	Wild oat	Gramineae
Brachiaria ramosa L. Stapf.	Signal grass	Gramineae
Cenchrus catharticus Dc.	Sandbur	Paniceae (Gramineae)
Chloris barbata Sw.	Finger grass	Chlorideae (Gramineae)
Cynodon dactylon (L) Pers.	Bermuda grass	Chlorideae (Gramineae)
Cyperus rotundus L.	Purple nut sedge	Cyperaceae
Cyperus esculentus L.	Yellow nut sedge	Cyperaceae
Digitaria sanguinalis (L.) Scop	Crab grass	Paniceae
Echinochloa colonum (L) Link.	Jungle rice	Gramineae
Echinochloa crus galli (L) Beauv.	Barnyard grass, Water grass	Gramineae
Eleusine indica (L.) Gaertn.	Goose grass	Chlorideae
Equisetum arvense	Horse tail	Equisetaceae
Eragrostis pilosa	Love grass	Festucaceae
Eriochloa gracilis H. B & K	Cup grass	Paniceae
Fimbristylis miliaceae	Hoorah grass	Cyperaceae
Lolium temulentum L.	Rye grass	Gramineae
Oryza sativa var *fatua*	Wild rice	Oryzeae (Gramineae)
Panicum fasciculatum Sw.	Brown top panicum	Paniceae (Panicum)
P. javanicum Poir	Panicum	Paniceae (Panicum)
P. prostratum Lank.	Panicum	Paniceae (Panicum)
P. ramosum L.	Panicum	Paniceae (Panicum)
P. repens L.	Panicum	Paniceae (Panicum)
Paspalum dilatatum Poir.	Dallies grass	Paniceae (Paspalum)
P. conjugatum Bergins	Sour grass	Paniceae (Paspalum)
P. disticum L.	Knot grass	Paniceae
Phalaris minor Retz.	Canary grass	Gramineae
Phragmites varka Retz.	Common reed	Gramineae
Poa annua L.	Annual blue grass	Gramineae
Saccharum spontaneum L.	Tiger grass Giant reed	Andropogoneae (Gramineae)
Sagittaria pygmacea	Arrow head	Alismafeceae
Scirpus articulatus	Bull rush	Cyperaceae
Setaria glauca (L) Beauv.	Yellow fox tail	Paniceae (Gramineae)
Sorghum halepense (L) Pers.	Johnson grass	Andropogoneae (Gramineae)

Control measure of quack grass

1. Herbicides like TCA at 6 kg/ha are useful to control this weed effectively on brassicae, legumes and carrot.
2. Dalapon (sodium) at 5 kg/ha and amino-triazoel at 4.5 kg/ha. may be sprayed on its active growing foliage for its effective control in legumes, carrot and orchards.
3. EPTC at 4.5 1/ha controls Agropyron in potatoes.
4. Atrazine 4.5 kg/ha. control this weed in maize.
5. Glyphosate 1.5 kg/ha when applied to actively growing foliage, is translocated to the rhizome and kills it.
6. Dichromate 5 kg/ha and Amitrole 1% spray are very effective.

2. *Andropogon annulatus*

Fox-tail grass...Andropogoneae
Padhar (Hindi)

Andropogon is a perennial pasture weed, also common in waste lands and road sides. The main stem is rhizomatous, remains underground, covered with scale leaves. The inter-nodes are purplish, smooth and glabrous; ligule membranous; leaf-blade is linear-lanceolate; spikes erect, stalk of inflorescence is long, slender; spillets sessile; stamens free; ovary with two feathery stigmas. It reproduces by seeds.

Control measures

Pre-emergence application of chlropropham (CIPC) at 3 to 4 kg/ha gives a good control of Andropogon.

Economic Importance: It produces considerable amount of fodder.

3. *Avena fatua*

Wild oat .... Gramineae

Wilt oat is troublesome and noxious winter annual, grassy weed, which mimics the cultivated oat in appearance. It infests mainly wheat, barley and oats.

The stems are upto 1 m high and leaves upto 20 cm long. The root system is fibrous; spikelets are long with dark awns, seeds are hairy and ripen earlier than wheat, barley and oat. Seeds are shattered at premature stage before the cultivated crops are harvested. Seeds remain viable in the soil for years; seed production capacity is also very high (about 5000 seeds per plant/year); its separation from cereal grains is very difficult and thus disseminated largely as seed contaminant.

Allelopathic effect: Wild oat residues inhibits germination and growth of wheat due to allelopathic mechanism.

Control measures

1. Since the main source of infestation is seed, straw and manure; clean seeds of grain crop should be used.
2. Rotation of crops with legumes or oil seeds is the best method of its control.
3. Weedicides like metoxuron 1 kg/ha as pre-emergence, can be incorporated in the soil upto 5 cm depth to control wild oat in beans, carrot and peas.
4. Barban 0.4 kg/ha as post-emergence should be sprayed at 2-3 leaf stage of wild oat in peas, beans and sugarbeet.
5. EPTC at 4.5 l/ha as pre-plating incorporation can give its good control in potato field.
6. Diallate 3 l/ha or tri-allate 1.5 l/ha as preplant incorporation to the soil, is beneficial.
7. Terbutryne 0.75 kg/ha sprayed at maximum tillering phase of wheat can kill 70% of wild oat, while 2,4-D at 0.5 kg/ha spray kills 40 per cent.

4. *Brachiaria remosa*

Signal grass.. Gramineae
Bhosa (Hindi)

Signal grass is an annual wild weed, usually found in cultivated fields of maize and sugarcane. Stem is 30 to 90 cm long; leaves 5 to 10 cm long, glabrous; sheaths pubescent; Spikelets sessile and pedicelled; glumes are membranous. It is propagated by seed. It has become an alternate host of cane mosaic virus.

Control measures

1. Hoeing and clean cultivation are essential to keep off the weed. In pigeonpea, two hand-weedings at 25 to 45 days after planting, following by pre-emergence application of alachlor (lasso) 1.5 kg/ha can keep the field weed-free.
2. Flumeturon at 2.5 kg/ha as pre-emergence can be applied, a day after sowing of cotton and as post-emergence spray at 6 weeks after sowing, to control this weed effective.
3. Simazine 1.75 kg/ha as pre-emergence application can control signal grass very effectively.
4. Oxadiazone 1 kg/ha, dichlormate 5 kg/ha and butachlor

(machete) 1 l/ha can control this weed is broadcasted rice.

5. *Cenchrus catharticus*

Sandbur..Paniceae

Sandbur is an annual herbaceous weed, common in cultivated fields, roadsides and pastures. Stem is of 50 to 60 cm height; root fibrous; leaves 25 to 30 cm long; leaf-sheath glabrous; inflorescence is a dense receme upto 10 cm long; fruit is a spherical bur. It reproduces by seeds.

Control measures

1. Deep tillage in the previous year or deep discing can keep this weed under control next year.
2. Propazine or atrazine 0.75 kg/ha each as pre-emergence application is very effective.

6. *Chloris barbata*

Finger grass .... Chlorideae

Finger grass is an annual herbaceous weed which is most common in road sides. Stem is erect, 50 to 100 cm high; roots fibrous; leave linear. Inflorescence is multiple spike, soft and silky white. Fruit is a caryopsis. It reproduces by seeds.

Control measures

1. Diuron 1 kg/ha and fluometuron 2.5 kg/ha as pre-emergence application, can control this weed in cotton.
2. Alachlor (lasso) 2 kg/ha as pre-emergence is effective.

7. *Cynodon dactylon*

Bermuda grass .... Chlorideae
Doob grass (Hindi)

It is a creeping perennial sod grass, a most trouble-some and pestiferous weed in cultivated lands. It is hard, pernicious, aggressive and invasive weed, found everywhere. It is found in sugarcane, rice, cotton and groundnut fields etc. It is very hardy and much branched with numerous stems below and above soil. Root is fibrous, creeping stems produce adventitious roots at each node. The underground stems are hard, brittle, with short scale leaves which occur mainly to furrow-slice depth of soil. Leaves are linear, 3 to 10 cm long; inflorescence consists of 4-7 auxillary spikes, originating from a common point and radiating in a cluster of 2 to 8 from the end of the culm. It grows all over the year and remains green. It reproduces by creeping rhizomes, stolons and also by seeds. Seed production is about 150 to 170 per plant. It is reported from Dharwar that the extent of yield reduction was 49% in cotton, 65% in wheat, due to infestation of Cynodon. It also depletes nutrients, moisture, soil aeration, light, space and also becomes alternate hosts of many pests and disease.

Allelopathic effect: Decayed residues inhibit seed germination, root and top growth of barley.

Control measure

1. It is found difficult to eradicate this noxious weed by manual, mechanical or chemical means and also costly. Hence an integrated approach is needed for its control.
2. Summer tillage by tractor ploughing is a normal practice to check its growth and spread.
3. Glyphosate 1.5 kg/ha is used as pre-plant in row crops. Dalapon or Amitrole 2 kg/ha each give a good control. Dalapon is applied twice, just after the rabi crop harvest and before monsoon to ensure its control.
4. As a pre-emergence, atrazine can also be used. Paraquat 1 kg/ha is applied 2 weeks before summer tillage to suppress it growth.
5. EPTC at 4 kg/ha as pre-emergence or chlorpropham 5 kg/ha can give a good control. However, a combination of MCPB (0.25 kg ai/ha) + MCPA (0.25 kg ai/ha) in proportion of 3:2 ratio is found to be the best.
6. For controlling Cynodon, TCA at 5.0 kg/ha as pre-planting, followed by dalapon at 1% spray is the best. Amitrole can also be applied as 1% liquid spray.
7. 2,4-D + dalapon at 2 kg/ha in 800 litre water/ha can be spread into the soil immediately after ploughing.
8. Simazine + atrazine at 2 kg/ha as pre-emergence application, followed by post – emergence of dalapon at 4.5 kg/ha give 90% control of Cynodon.

Economic importance: Bermuda grass is best for lawn and a soil binding material for bunds and check soil erosion. It is a good drought tolerant pasture grass used as good green fodder for horse, goat, and rabbit. Stalk juice is applied to bleeding cuts and wounds; chewed while paining the teeth; and used as diuretic and urinary sedative.

8. *Digitaria sanguinalis*

Crab grass ....Paniceae
Siur (Hindi)

Digitaria is an annual herbaceous grass, common in cultivated fields, roads, ditches, pastures and especially on light soils. It is a slender grass; root is fibrous; stem is semi-erect and branches profusely at the base, grows 50 to 100 cm tall; leaves are pubescent, 5 to 10 cm long; ligules with long silky hairs; inflorescence is composed of multiple spikes. It reproduces both by seeds and vegetatively, Vengris *et al.*, (1953) reported the major nutrient removal by crab grass is very high. (N = 2.0%, P_2O_5 = 3.36%, K_2O = 3.48%, Ca = 0.27%, Mg = 0.54%) and thus it becomes a great competitor for nutrients with crop plants.

Allelopathic effect: Its decaying roots inhibit root growth of maize and yield of cotton and groundnut.

Control measures

1. Clear cultivation like hand-weeding and hoeing is normally practiced. Hand weeding is found to be better in ragi than selective herbicides.
2. Simazine 2 kg/ha, MSMA 2 kg/ha or matribuzine 1 kg/ha is found to be effective to control crab grass.
3. Other effective weedicides are: bromacil 2 kg, diuron 4 kg, monuron 3 kg, linuron 1.5 kg, PROBE 2 kg, nitralin 1 kg, Promatryne 2 kg, trifluralin 1.5 kg, PCP 5 kg, DSMA 2 kg and dalapon 4 kg/ha.

Economic importance: it is good as green fodder.

9.(a) *Echinochloa colonum*

Jungle rice .... gramineae
Sawan (Hindi)

Jungle rice is an annual, herbaceous, tuftad grass of the rainy season, commonly found in mixed population among grassy and broad-leaf weeds, particularly in rice, throughout the season; waste places and among ditches. It grows very fast and establishes well in moist soils. It resembles rice and mimes rice plants and thus is found very difficult to identify during its growth period. It grows in rice nursery, naturally mixed with paddy and multiplied in situ in large numbers. Roots are fibrous; stem decumbent; tillering profusely from the base; leaves 5 to 10 cm long; inflorescence is a purple panicle. It is propagated only by seeds.

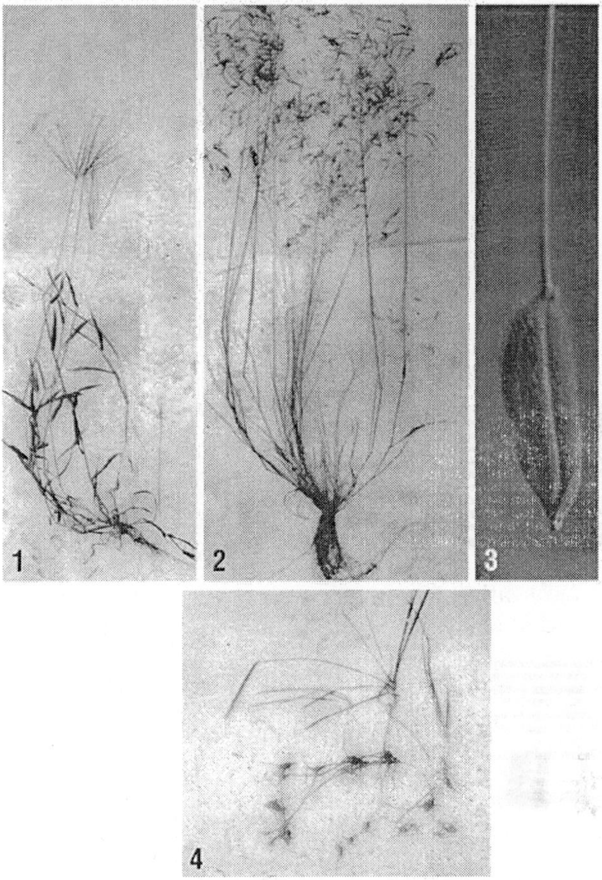

Plate 1: (1) *Digitaria sanguinlis* (Crab grass), (2) *Eragrostis tremi* (Love grass), (3) *Oryza sativa* var. fatua (Wild rice), (4) *Cyperus rotundus* (Purple nutsedge)

Control measures

1. Mechanical method may be practiced according to availability of labour.
2. Propanil 2.5 kg ai/ha, applied 2 weeks after sowing paddy in the nursery bed, can kill 70% of jungle rice.
3. In direct-seeded wet-land rice, machete (g) at 20 kg/ha, benthiocarb (g) at 20 kg/ha or molinate 2 kg/ha are very effective.

4. Pre-emergence application of simazine plus atrazine at 1 kg/ha each followed by a post-emergence application of 2,4-D at 4.5 kg/ha can give as high as 90% control of jungle rice.
5. Mixture of machete + carbofuran insecticide granules, is most effective in killing grasses, sedges, broad-leaf weeds and also stem borer in rice. Machete mixed with phorate granules can control all kind of weeds and leaf-hopper also. Such combination of herbicide + insecticide is quite compatible and also economic to apply.

Economic importance: It is a good fodder grass.

9. (b) *Echinochloa crusgalli*

Barnyard grass, water grass .... Gramineae
Sama ghas (Hindi)

Barnyard grass is the most troublesome weed in low-land rice, particularly in Kharif season during June to September and causes a substantial reduction in yield. Like jungle rice, it also mimics rice plants and multiplies along with transplanted rice seedlings which are distinguishable during its vegetative growth period. It reduces the grain yield as high as 35 per cent.

Stems are erect, 30 to 70 cm long; tillers profusely from the base; roots are fibrous; spikelets are awned about 3.5 mm. It differs from rice since it has no auricle or ligule. It is propagated only by seeds; these seeds are mixed with paddy seeds for easy propagation. Barnyard grass seeds can germinate even from 10 cm depth of soil.

Allelopathic effect: Its residues inhibit hypocotyle growth of soybean upto 50%.

Control measures

1. Hand-pulling of weeds in the nursery, two to three times, is a normal practice.
2. Among weedicides, propanil 2.5 kg ai/ha applied at 2 to 3 leaf stage, is the best and hence commercially used for its control.
3. Machete 2 l/ha randox (CDAA) 4 l/ha, cobex (dinitramine) 0.5 kg/ha or nitraline 0.5 kg/ha can control water grass effectively.
4. Alachlor (lasso) 2 kg/ha as pre-emergence application, followed by 2,4-D at 1.5 kg/ha at full-tillering stage of rice, can keep the plot weed-free for 30 o 35 days.

10. ***Eleusine indica***

Goose grass .... Chlorideae

Goose grass is an annual herbaceous grass, commonly found in cultivated fields (wheat, maize, ragi, rice); road sides and pastures. It is heliophytic and hence suppressed by reduction in sunlight intensity.

Stem is 30 to 100 cm tall and highly branched, fibrous root; leaves are linear-lanceolate 20 to 30 cm long; inflorescence is composed of multiple spikes (4 to 8). It reproduces by seeds.

Allelopathic effect: Decomposing residue inhibits seed germination of growth of many spices.

Control measures

1. Clean cultivation by hoeing and weeding is the best.
2. Pre-emergence application of atrazine 1 kg/ha mixed with alachlor 1.25 kg/ha is quite effective.
3. Machete (g) 20 kg, Propanil 5 kg, bentiocarb 2 1, dichlormate 5 kg or fluchloralin 1 kg ai/ha are very effective to control goose grass.
4. Pre-emergence application of alachlor 1.5 kg/ha, followed by 2,4-D at 1.5 kg/ha spray is very effective.

11. ***Eragrostis pilosa***

Love grass ....Festucaceae
Koos (Hindi)

It is a perennial stout grass, found in sandy dry areas, grasslands and road side; leaves are 40 to 50 cm long, erect, margins minutely scabrid; spikelets borne direct on the main rachis above. It is propagated by seed and root stock.

Control measures

Deep plaughing mowing repeatedly at its early growth can control this weed.

Economic importance: It is used for making ropes and also considered sacred among Hindus. A ring of it is worn when performing religious ceremonies.

12. ***Lolium temulentum***

Rye grass .... Gramineae

Rye grass is a comparatively new annual weed species,

commonly grown in sub-tropical regions in the wheat field. In the admixture of wheat seeds, this seed can easily be recognized because of its shape, size, seed surface and colour.

Roots are fibrous, stems erect, usually branched at the base; clums cylindrical; leaves solitary at the nodes, consisting of sheath, ligule and blade; lemmas are membranous and shining entire, 2 to 5 toothed at the apex, tightly enclosing one another.

Allelopathic effect: Decomposing residues of rye grass are toxic to seedling growth of oat, lettuce and *Trifolium* spp.

Control measures

1. Herbicides like alachlor 2 kg/ha alone or alachlor followed by 2,4-D at 2 kg/ha, applied 30 days after sowing of wheat is very effective and thereby wheat yield can be increased by 30% but it is slightly toxic to wheat.
2. Dichlorprop (Weedone) or 2,4-DP at 1.0 kg/ha as post-emergence is very effective to control ryegrass.

13. *Oryza sativa* var. *fatua*

Wild rice .... Oryzeae

Wild rice is a troublesome weed in rice growing areas. It mimics cultivated rice during its vegetative phase. It matures and shatters seeds earlier, before the cultivated rice crop ripens in the main field and germinates only in wet soils, till then it remains dormant in the field.

Root is fibrous; stem erect, 75-150 cm long and tillers at the base. Leaves are linear lanceolate; inflorescence is an open, drooping panicle and most of the spikelets have long awns which is the peculiar characteristics. It reproduces only by seeds and disseminates largely by seed contaminants.

Control measures

Since wild rice flowers earlier than medium duration rice, it can be cut and removed from the field before the panicles are matured.

14. *Panicum repens*

Torpedo grass ....Paniceae

Penicum spp. is mostly a perennial grass; grows stiff, erect in clusters on bunds, side of tanks and wet places. Underground stems are thick and hard which are pointed a the end; aerial stems are covered with scale leaves at each node, leaf sheath is smooth, glabrous; lower nodes bear adventitious roots; leaf margin is minutely serrate; legule consists of a short thin membrane, hairy at the top. Inflorescence is much branched; a compound spike varying in length; spikes are many; spikelets are turgid, sessile and densely packed.

Control measures

1. Once Panicum is established, it is difficult to eradicate, particularly Panicum repens. It can be destroyed by digging out the underground rhizomes.
2. Combination of bromacil 4 kg/ha mixed with terbacil 4 kg/ha or glyphosate 4 l/ha or dalapon 4 kg/ha can give its effective control.
3. Dalapon 5 kg/ha + glyphosate 4 l/ha is the best to control this grass.

15. *Paspalum* spp.

Knot grass, dallies grass, sour grass..........................Paniceae

Paspalum is a perennial sod-forming grass which is common in pastures, road sides and perennial crop areas.

Root is fibrous, stems are tufted on very short rhizomes, ascending, stoloniferous; nodes prominent and hairy; leaves are linear lanceolate, 8 to 12 cm long. Inflorescence consists of two thin terminal racemes, rachis is flattened and winged. It reproduces by both seeds and stolons.

Control measures

1. ***Panicum distichum*:** Dalapon 6 kg plus amino-triazole (ATA) 4 kg/ha or 2,4-D at 1 kg/ha or DSMA at 3 kg/ha can give its complete control within 75 days.
2. ***Panicum conjugatum*:** (i) Glypohsate 0.8 kg/ha is most effective against it. (ii) Paraquat 0.4 kg/ha + diuron 0.5 kg/ha is also effective.

16. *Phalaris minor*

Canary grass .... Gramineae

Canary grass is an annual weed, mostly found in wheat and gram growing areas, which is spread through seed contamination by the seed-supply sources. It sets seeds almost the same time as the wheat crop and also grows at the same height and hence easily contaminated at crop harvest. Each canary grass produces 200 to

500 shiny, small black seeds which can easily be identified by its fox-tail type spike.

Control measures

1. Canary grass seeds can easily be sieved from wheat grains and clean seeds can be used.
2. Crop rotation with legumes and oilseeds will help to control this weed.
3. Spraying of TOK-E 25 at l/ha can give 90% control of this weed.
4. Fluchloralin 2 kg/ha or terbutryn 0.15 kg/ha as pre-emergence, sprayed at maximum tillering stage of wheat, cat give about 70% control of canary grass.
5. Nitrofen 1.5 kg/ha, alachlor 2 kg/ha as pre-emergence application can control this weed effectively.

17. *Poa annua*

Blue grass, wild buck wheat .... Gramineae

Poa annua is an annual grass, commonly fond in cultivated fields of Potato, radish, turnip carrot etc. It belongs to the sub-tribes of Festuceae of the family Gramineae. It has got all the characteristics of Gramineae. The special characteristics is that the spikelets are arranged in solitary spikes, the lower glume suppressed, except in the terminal spikelet.

Control measures

1. Dalapon 5 kg, TCA 10 kg, trifluralin 1 kg or diallate 1.5 kg/ha as preplanting, can control this weed effectively.
2. Iso-proturon 1 kg/ha or metoxuron 2 kg/ha as pre-emergence, is proved to be effective.

18. *Saccharum spontaneum*

Tiger grass ....Andropogoneae
Kans (Hindi) or wild reedcane

'Kans' is a deep-rooted, Coarse, fast-spreading-perennial grass, found mostly in river-belt and low-lying grounds. It is a pernicious weed which reproduces rapidly and tillers profusely. It becomes aggressive on well drained, moisture retentive soils, while susceptible to water-logging. It is a noxious weed.

Rhizomes found up to 30 cm depth of soil; roots are creeping in nature, forming a continuous net work throughout the land; stem erect, 2 to 4 m high, leaves narrow panicles 25 to 30 cm long. It is propagated both by seeds and underground rhizomes.

Control measures

1. Deep and continuous cultivation, flooding the infested land; smothering by growing vigorous rainy season crops, constant cutting of aerial shoots are the common practice for suppressing the spread of tiger grass.
2. Above physical control measures can be followed by application of TCA (sodium) at 30 kg/ha to reduce its population to a minimum.
3. Spray of monuron 5 kg/ha can prevent dissemination of wind-borne seeds of S. spontaneum.
4. Dalapon 20 kg/ha + amino-triazole 10 kg/ha can control regenerated tiger grass.

19. *Setaria glauca*

Yellow foxtail .... Paniceae

Yellow foxtail is an annual noxious weed, found in fields of plantation crops, pastures and road sides. It reproduces rapidly and tillers profusely. Seeds disseminate with wind and germinate all over the field soon after pre-monsoon showers.

Root is fibrous; stem decumbent, 30 to 80 cm tall; leaves are linear-lanceoalte, 15 to 25 cm long, leaf base pubescent. Inflorescence is a dense, bristly panicle, 10 to 20 cm long. The stiff erect culms do not usually branch. Spikelets are numerous and closely set along the rachis of inflorescence which are the special characteristics for its identification. It is reproduced by seeds are regenerate from old culms.

Allelopathic effect: Interference by yellow foxtail affects crop yields, particularly of maize, soybean and sorghum.

Control measures

1. Young weed seedlings can be removed by hand pulling during rainy season from different patches.
2. Dichlormate 3 kg/ha along with high does of nitrogenous fertilizers can be applied for its control.
3. 2,4-D at 1 kg/ha plus diuron 0.5 kg/ha spray is very effective to control this weed.
4. Glyphosate 2 kg/ha as post-emergence at its active growing stage can arrest the growth effectively as it translocates to the root clumps and kills it. This herbicide is found to be the best to control this weed.

5. Effective herbicides to control yellow foxtail are; dalapon 5 kg, machete 21, randox 2 kg, naptalam 4 kg, nitralin 1 kg, trifluralin 1 kg, PCP 5 kg, DSMA 2 kg, MSMA 2 kg, chloramben 3 kg, benthiocarb 2 kg, diuron 4 kg and chlorxuron 3 kg/ha.

20. *Sorghum halepense*

Johnson grass ....Andropogoneae

Johnson grass is a stout, robust, persistent, perennial noxious weed commonly found in cultivated fields of maize, millets, sugarcane cotton etc. in orchards pastures and road sides during kharif season. It is a deep-rooted weed, rhizomes tend to grow chiefly downward and spread through stout rhizomes and also reproduces by seeds. Seeds are produced in winter season and easily disseminated with wind and are germinated densely in moist soil, soon after the first showers. Seedlings form rhizomes within 3 to 4 weeks and deep-rooted and hence difficult to eradicate when firmly established.

It is poisonous as animal feed at tillering phase, because of high prussic acid, (HCN) content. Primary root stocks are alive in the ground, stems produce adventitious roots from lower nodes, stands erect, 90 to 120 cm high, leaves are linear-lanceolate, glabrous, dark-green, 50 o 75 cm long and 3 to 4 cm wide. Inflorescence is an open panicle, 20 to 40 cm long. Seeds oblong, lanceolate.

Allelopathic effect: Interference by Johnson grass can reduce soybean yield by 40% and decaying underground organ can inhibit root and shoot growth of barley.

Control measures

1. Cultural methods fail to control this persistent weed when established but pre-planting application of basalin (fluchloralin) 2 kg/ha can control this weed effectively.
2. TCA 60 kg/ha in conjunction with 5 to 6 cultivations can give it complete control (upto 90%), if applied just before the break of monsoon. The chemical is absorbed by rhizomes and thus killed almost completely.
3. Trifluralin 1 kg/ha and nitralin 3 kg/ha can control it effectively.
4. Dalapon 1 % or 10 kg/ha spray on established weed, is very effective.
5. Other effective weedicides are; glyphosate 5 kg, butachlor 2 kg, bromacil 5 kg/ha etc.

B. Sedges – Cyperaceae

21. *Cyperus esculentus*

Yellow nut sedge .... Cyperaceae

Yellow nut sedge is a perennial sedge, commonly found in cultivated lowlands. It has become a serious range weed, infesting grain crops, vegetables, fruit crops, lawn and ornamental plants.

The leaves are erect, linear; inflorescence is an umbel of multiple pedicelled spikelets. Tubers are under ground at the base of the plant, basal bulb is first formed and then new shoots and rhizomes are developed which accumulate the nutritious substance. They reproduce both by seeds and tubers. Tubers have varying degrees of dormancy and these are also having drought-resistant properties and hence it is very difficult to kill them.

Allelopathic effect: Its allelopathic effect is observed on soyabean, maize and oats.

Control measures

1. Integrated control measures like deep summer tillage, competitive cropping and use of suitable herbicides are helpful to suppress yellow nut sedge.
2. Herbicidal control is not so successful because of poor translocation and dormant nature of tubers. Among the herbicides, glyphosate 2 kg/ha as post-emergence application at 5 to 7 leaf stage; followed by application of dalapon 200 ppm cause a complete kill of underground parts.
3. 2,4-D at 2 kg/ha plus 2,4,5-T at 2.5 kg/ha mixed with 5% Jaggery solution as pre-emergence application, cause 90% mortality and also suppress regeneration.
4. Amitrol 4.5 kg/ha + atrazine 1.5 kg/ha can control Cyperus effectively, causing 95 to 100% mortality. Similarly, 2,4-D at 2 kg/ha combined with atrazine 2 kg/ha can cause 80% mortality.

22. *Cyperus rotundus*

Purple nut sedge .... Cyperaceae
'Motha' (Hindi)

Purple nut sedge is a most persistent deep rooted, perennial, cosmopolitan, pestiferous sedge; most noxious of all tropical and sub-tropical weeds, particularly in the cultivated fields, lands and orchards. It becomes a great menace in all kharif season crops like maize, sugarcane, rice, groundnut etc. and exhibits no preference for soils.

Rocheconste (1956) estimated its removal of 2.17% N, 0.26% P_2O_5 and 2.73% K_2O from the soil. Bhardwaj *et al.*, (1968) estimated its NPK uptake upto 95.6 kg, 116 kg and 42.3 kg/ha respectively during Kharif season (April to October).

The rhizomes produces numerous chains of bulbs (nuts) or nutlets. Under favourable temperature and moisture conditions, single nutgrass tuber can develop as many as 100 tubers in 4 months. Its extreme prolific capacity due to underground rhizomes and tubers, has created a challenging problem.

A nut or tuber possesses scaly leaves, inconspicuous nodes and internodes and minute bud which give rise to new shoots, rhizomes and roots. The basal bulbs contain stored food. A tuber produces a few rhizomes with basal bulbs and the net work consists of 4 to 5 dormant tubers for each 'active' one; and the sequence is repeated, forming a big chain of a dozen inter-connected tubers, some of which descend to considerable depth of the soil. This nutgrass thus grows luxuriantly even during dry months because of the presence of well-developed storage organs. The first tuber formation is completed within 10 to 15 days after germination. The tuber changes its colour from brown to black with age. It is estimated that about 50% of tubers are located in the top 15 cm depth, 45% in 15 to 30 cm depth and remaining 5% descend beyond.

Its stem is triangular, 15 to 50 cm tall without nodes, glabrous and green; leaves are numerous, linear, dark green, 5 to 15 cm long; inflorescence is a compound umbel, supported by bract of leaves. Since the seed viability is low, its propagation is mainly through tubers (nuts).

Allelopathic effect: Soils previously infested with nutsedge reduce germination of mustard, barley, cotton and yield of rice.

Control measures

1. Integrated control measure is essential. Cultural practices like deep summer tillage by tractor to expose the tubers to the sun; continuous weeding for 3 years so as to prevent the aerial shots to come up; growing green manure crops, and suppress this weed growth. Exposing of rhizomes and tubers to the sun create exhaustion of starch and thereby death occurs. Sinha and Thakur (1967) reported that continuous cultivation in 2 years at an interval of 3 weeks can control nutgrass population as high as 99 per cent. It is observed that tubers at 30 cm depth germinate almost cent per cent; but below 60 to 90 cm, tuber

germination is practically nil.

2. Glyphosate 2 kg/ha as post-emergence can arrest the growth of nut sedge completely; while dalapon 4 kg/ha or 200 ppm brings about complete scorch and desiccation and thereby kills underground parts of nutsedge.
3. 2,4-D (sodium) a 3 kg/ha (2000 ppm) can kill 80 to 90% nut sedge; while a mixture of 2,4-D (amine) 0.5 kg/ha plus MSMA 0.5 kg/ha can kill the basal bulbs and attached tubers.
4. Nutsedge is susceptible to a mixture of amitrole 4.5 kg/ha or 1% plus atrazine 2 kg/ha.
5. EPTAM at 7.5 l/ha as post-emergence, followed by dalapon 8.5 kg/ha can kill nutsedge almost completely.
6. Bromacil 2 kg/ha or Sinbar 2 kg/ha can control top growth of nut sedge for at least 4 months.
7. Dicamba 1 kg + 2,4-D at 2 kg/ha added with molasses at 3% spray can easily be translocated from mother to daughter tubers and kill them completely.
8. EPTC at 6.5 kg/ha can be incorporated in the soil to check this weed for 12 to 15 weeks and thus the crop can be saved.

Economic importance: It has medicinal uses. It is applied externally on breast, it stimulates secretion of milk. Essential oil is extracted from tubers, fragrant agarbati sticks are also prepared using its tubers.

23. *Fimbristylis annua*

Hoorah grass .... Cyperaceae

Hoorah grass is an annual sedge, commonly found in low-land rice fields, swampy areas and river banks.

Roots are fibrous, stem angular and erect, 30 to 50 cm tall and highly branched; leaves linear; inflorescence is a cluster of spikelets, arranged in a compound umbel. It reproduces by seeds.

Control measures

1. Weed-free condition upto three weeks from planting of rice either manually or mechanically, can give an economic yield.
2. Machete(g) 20 kg/ha can kill the cyperous weed very effectively. It can also be mixed with carbofuran to kill stem borer; and mixed with phorate to kill the leaf hopper.
3. Stam F-34 at 5 l/ha (actual) or Oxadiazon 1.25 kg/ha is also effective.

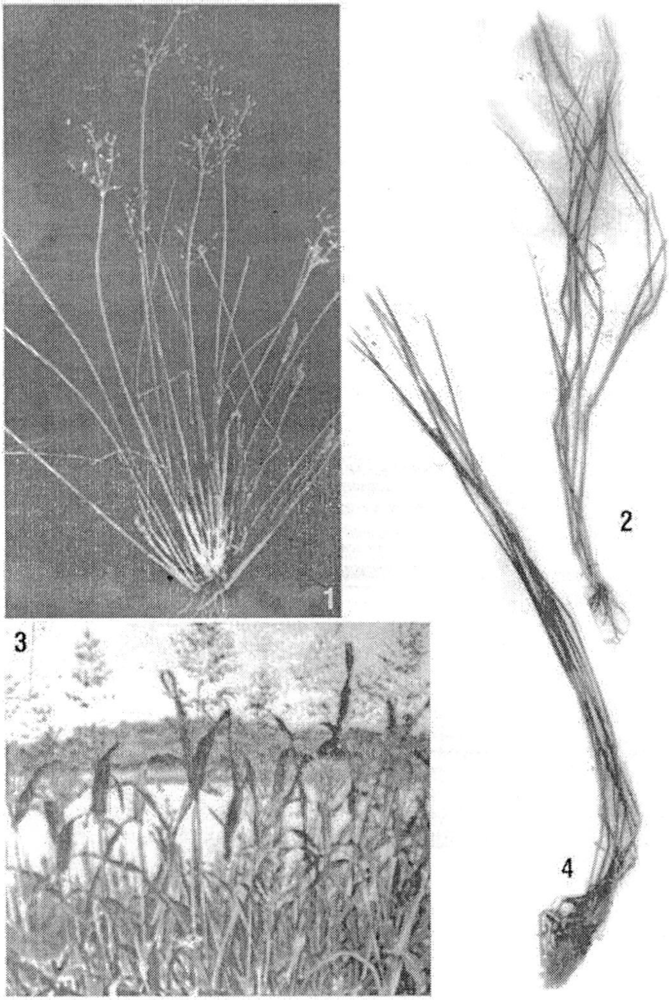

Plate 2: (1) *Fimbristylis annua* (Hoorah grass), (2) *Setaria glauca* (Yellow fox tail), (3) *Scripus* sp. (Bul rush), (4) *Sorghum halepense.*

C. Liliaceae

24. *Allium* spp.

Wild onion and Wild garlic ....Liliaceae

Wild onion (*A. canadense*) and Wild garlic (*A. vineale*) tubular leaves are round, arranged at the base, long, fleshy and upright which are found in open situation. Root stocks form a bulb; flowers bisexual, arranged in umbel-like heads. They are disseminated with seeds of small grains and spread over the fields. Their seeds mature at the same time as those of winter grains and hence easily contaminated with crop seeds at harvest. It is a troublesome weed, causing tainting of milk and flour.

Control measures

1. It can be controlled by rotational cropping and also discontinuing grain crop cultivation for a year or two.
2. Those weeds are also disturbed by cultural practices adopted to destroy young ones.
3. Aminotriazole 6 kg/ha applied before flowering can give a good control.

Monocotyledonus seeds (Family-wise) (Grasses, sedges, Liliazeae and others)

Family	Scientific name	Common name
A. GRASSES (GRAMINEAE)		
Agrostideae	*Agrostis palustris* (Huds) Pers.	Bent grass
	Aristida dichotoma Mitchx.	Threeown
Andropogoneae	*Andropogon annulatus* Forsk.	Foxtail grass
	Saccharum sontaneum L.	Tiber grass
	Sorghum halepense (L) Pers.	Johnson grass
Chlorideae	*Chloris barbata* Sw.	Finger grass
	Cynodon dactydon (L) Pers.	Bermuda grass
	Eleusine indica (L) Gaertn.	Goose grass
Festucaceae	*Eragrostis pilosa*	Love grass
Gramineae	*Agropyron repens* (L) Beauv.	Quack grass
	Avena fatus L.	Wild oat
	Brachiaria remosa (L) Stapf.	Signal grass
	Echinochloa colonum (L) Link.	Jungle rice
	Echinochloa crusgalli (L) Beauv.	Barynyard grass
	Lolium temulentum L.	Rye grass
	Phalaris minor Retz.	Canary grass
	Phragmites varka Retz.	Columon recd
	Poa annua L.	Blue grass
Paniceae	*Cenchrus cathurticus* Del.	Sand bur
	Digitaria sanguinalis (L) scope.	Crab grass
	Eriochloa gracilis H. B & K	Cup grass

Family	Scientific name	Common name
	Panicum spp.	Panicum
	Paspalum spp.	Knot grass
	Setaria glauca (L) Beauv.	Yellow fox tail
Oryzeae	*Oryza sativa* var. *fatua*	Wild rice
B. SEDGES		
Cyperaceae	*Cyperus esculantus* L.	Yellow nut sedge
	Cyperus rotundus L.	Pumple nut sedge
	Fimbristylis miliacea	Hoorah grass
	Scirpus articulatus	Bull rush
C. LILIACEAE		
	Allium canadense L.	Wild onion
	Allium vineale L.	Wild garlic
	Asphodelus fistulosus L.	Jungle onion
D. ALIAMATECEAE		
	Sagittaria pygmacea	Arrow head

Herbicides for Monocot Weeds

Weed species	Effective herbicides (kg or l/ha)
Agropyron repens	TCA (6), dalapon (5), EPTC (4), atrazine (4) glypohsate (1.5), dichlormate (5)
Allium spp.	Aminotriazole (6)
Andropogon annulatus	CIPC (3)
Avena fatua	Metoxuron (1), barban (0.4), EPTC (4), diallate (4), terbutryne (0.75)
Brachiaria remosa	Flumeturon (2.5), Simazine (1.75)
Cenchrus catharticus	Propazine or atrazine (0.75)
Chloris barbata	Diuron (1), fluometuron (2.5), alachlor (2)
Cynodon dactylon	Glypohsate (1.5), dalapon (2), EPTC (4), MCPA + MCPB (0.25 each), TCA (5)
Cyperus esculantus	Glyphosate (2), dalapon (2), 2,4-D + 2,4,5-T (2.5 each, amitrole (4.5) + atrazine (1.5).
Cyperus rotundus	Glyphosate (2), dalapon (4), amitrol (4.5), EPTAM (7.5), EPTC (6), bromacil (2), dicamba (1) + 2,4-D (2).
Digitaria sanguinalis	Simiazinc (2), MSMA (2), metribuzin (1)
Echinochloa colonum	Propanil (25), machete (gr) (20), bentio (gr) (20), molinate (2)
Echinochloa crusgalli	Propanil (2.5), machete (2), Randox (4)
Elebsine indica	Atrazine (1), alachlor (1.25) machete gr. (20), Propanil (5),

Weed species	Effective herbicides (kg or l/ha)
	benthiocarb (2)
Fimbristylis annua	Machete gr (20), stam F-34 (5), Oxadiazin (1.25)
Lolium temulentum	Alachlor (2), dichlorprop (1)
Panicum repens	Bromacil (4) + terbacil (4), glyphosate (4), dalapon (5)
Paspalum spp.	Dalapon (6), glyphosate (0.8), diuron (0.5)
Phalaris minor	Toke-25 (2), fluchloralin (2), nitrogen (1.5).
Poa annua	Dalapon (5), TCA (10).
Saecharum spontaneum	TCA (30), monuron (5), dalapon (20) + amino-triazole (10).
Setaria glauca	Glyphosate (2), 2-4-D (1) + diuron (0.5), dichlormate (3).
Sorghum halepense	TCA (60), trifluralin (1), nitralin (3), dalapon (10), glypohsate (5), bromacil (5), butachlor (2)

CHAPTER 3

Dicot Weeds

The embryo of dicotyledons has two cotyledons, stems contain open vascular bundles usually in one ring, leaf net-veined, flowers often with distinct calyx and corolla, usually pentamerous or tetramerous.

1. *Acalypha indica*

Copper leaf .... Euphorbiaceae
Kuppi (Hindi)

Acalypha is an annual herbaceous weed, found almost every where including cultivated fields of upland rice, onion, maize etc.

It has a tap root and an erect branching stem, 30 to 50 cm tall which produces milky juice. Leaves alternate, long petioled, obvate with serrate margins; inflorescence is an auxillary spike, top of which bears male flowers while in the lower part, female flowers are produced. Fruit is a capsule concealed by the bracts. It is propagated by seeds.

Control measures

1. Clean cultivation and weeding before seeding; use of hand hoe and rotatary weeder should be done.
2. In rice, one weeding at 3 weeks, followed by spraying of stem F-34 at 2.5 l/ha at 6 weeks is proved effective.
3. In onion, flucloralin (Basalin) 1 kg ai/ha at pre-plant, followed by one weeding is the best to control this weed.
4. In maize, pre-emergence application of atrazine 2 kg ai/ha or soil incorporation of EPTC at 3 kg/ha is the best for its control.

Economic importance: It is a medicinal plant.

Dicotyledonous weeds (alphabetic order)

Scientific name	Common name	Local name	Family
Acalypha indica Linn.	Copper leaf	Kuppi	Euphorbiaceae
Acanthosperum hispidum DC.	Bristly Starbur		Compositae

Scientific name	Common name	Local name	Family
Achyranthus aspera L.	Prickly chaff flower	Chirchire	Amaranthaceae
Amaranthus spionosus L.	Spiny pigweed	Katili chauri	Amaranthaceae
Amaranthus viridis L.	Slender pigweed	Jungle chacori	Amaranthaceae
Ambrosia artemistifolia L.	Common ragweed	-	Compositae
Anagallis arvensis L.	Scarlet pimpernel	Krishnaneel	Primulaceae
Argemone maxicana	Mexican prickly poppy	Goatweed, Kateli	Papaveraceae
Asteracantha longifolia L.	Creeping spiderling	Satyanashi	Acanthaceae
Boerhavia diffusa L.	Creeping spiderling	Bis-khapra	Nyctaginaceae
Brassica vaber DC.	Wild mustard		Cruciferеae
Calotrpis gigantean Br.	Giant milkweed	akona	Asclepidaceae
Carthamus oxyacantha Bieb.	Wild safflower	Pohli	Compositae
Cassia occidentalis L.	Coffee sena	Chakora	Caesalpiniaceae
Cassia tora L.	Sickle pod	Chakora-chakunda	Caesalpiniaceae
Celosia argentea Linn.	White cockscomb.	Safed Murgaka phool	Amaranthaceae
Chenopodium album L.	Lambsquarters	Bathua	Chino-podiacea
Chenopodium murale L.	Goose foot		Chenopodiaceae
Chorispora tenella L.	Blue mujstard		Crucifereae
Cirsium arvense L.	Canada thistle	Kateli, Kantaila	Compositae
Cleome viscos L.	Spider flower	Jungle hurhur	Capparidaceae
Clitoria ternate			Papilionaceae
Soccinia indica			Cucurbitaceae
Convolvulus	Field bind weed	Hirankhuri	Convolvulaceae
Corchorus acutangulus Lin.	pod rush	Jangli jute	Tiliaceae
Cucumis trigonus roxb.	Wild cucumber		Cucurbitaceae
Cuscuta spp.	Dodder	Akashbel	Convolvulaceae
Datura stramonium	Thornapple	Datura	Solanaceae

Scientific name	Common name	Local name	Family
Daucus carota	Wild carrot		Umbelliferae
Digera arvensis Forsk.	Amaranthus	Karigandhari	Amaranthaceae
Eupatorium odoratum	Spurge	Communist weed	Compositae
Euphorbia hirta L.	Garden spurge	Bandudhi	Euphorbiaceae
E. supine	Spotted spurge		Euphorbiaceae
Gynondropsis gynandra L.	Wild spider flower	Hurhur	Capparidaceae
Hibiscus vitifolicus			Malvaceae
Indigofera glandulosa Wild.	Wild indigo	Neel	Caesalpiniodea
Ipomea purpurea Lamk.	Morning glory	Karni	Convolvulaceae
Justicia prostrate Gamble.	Rushes	Derua	Acanthaceae
Lagasca mollis	Skeleton weed		Compositae
Lantana camara	White sedge, brush	Phulbakri	Verbenaceae
Lathyrus aphaca L.		Jungle Gobhi	Papilionaceae
Lebcus aspara		Guma	Labiatae
Loranthus sp.	Loranthus		Loranthaceae
Melilotus sp.	White sweet clover	Bonmethi	Papilionaceae
Melilotus indica	Yellow sweet clover	Pilasenji	Papilionaceae
Mimosa pudica L.	Sensitive plant, touch-me-not	Lajbanti	Mimosaceae
Mollugo vericillata	Carpet weed		Molluginaceae
Opuntia dillenii Haw	Pricklypear		Cactaceae
Orobanche aegyptiaca Pers.	Broom rape	Hadda, Banda	Orobanchaceae
Oxalis acetosella	Wood soral		Oxalidaceae
Parthenium hysterophorus L.	Ragweed	Congress weed	Compositae
Phyllanthus niruri L.	Gripe weed	Hazardana	Euphorbiaceae
Physalis minima L.	Hogweed, ground cherry	Banmakoa	Solanaceae
Pluchea lanceolata Oliv.	Arrowwod		Compositae
Polygonum plebejum	Smartweed		Polygonaceae

Scientific name	Common name	Local name	Family
Portulaca oleracea L.	Common purselane	Lunuk	Portulacaceae
Raphanus raphanisturm L.	Wild radish		Crucifereae
Rhus radicans L.	Poison ivy		Anacardiaceae
Salsola kali L.	Russian thistle		Chenopodiaceae
Solanum xanthocarpum L.	Night shade	Kateli	Solanaceae
Sonchus arvensis L.	Sowthistle		Compositae
Spergula arvensis L.	Cornspurry	Bandhania	Caryophyllaceae
Stellaria media (L.) vill.	Chickweed		Caryophyllaceae
Striga asiatica Lour.	Figwort, witchweed	Agiakhar	Scrophulariaceae
Tephrosia purpurea Pers.	Mild indigo	Sparphuka	Leguminoceae
Trianthema monogyma L. syn. *Portulacastrum* L.	Horse purselane	Patharchata	Aizoaceae
Tribulus terrestris L.	Punchure vine		Zygophyllaceae
Tridex procumbense L.	Tridex	Tunaki	Compositae
Vicia sativa L.	Common vetch	Ankta	Papilionaceae
Xanthium strumarium B.	Cocklebur	Chhotagokhru	Compositae

2. *Acanthosperum hispidum*

Bristly starbur .... Compositae

Acanthosprum is an annual herbaceous plant introduced from South America, found mostly in pastures, roadsides and cultivated fields of wheat, ragi, sugarcane, groundnut etc. This weed is recognized by forked branches bearing opposite leaves.

It has taproot; erect, diffusely branching stem, 50 to 100 cm tall, leaves are simple, sessile, opposite and irregularly toothed, inflorescence axillary; flowers crouded into heads, 5 to 8 female florets while disc florates are male. Fruit is short hooked, spiny achene with two terminal beaks; seeds without endosperm. It is reproduced by seeds. It is a host of number of viruses which attack cultivated crops.

Plate 3: (1) *Acalypha indica* (Copper leaf), (2) *Achantho sporium hispidum* (Starbur), (3) *Amaranthus viridis* (Rig weed), *Amaranthus spinosus* (Spiny pig weed), (4) *Gynandropsis gynandra* (Spider flower).

Control measures

1. Inter-cropping system is row-crops like pigeonpea with sorghum, cowpea etc. can keep crop-weed balance and do not influence weed infestation.
2. In wheat, fluchloralin or metoxuron at 1.0 kg ai/ha is effective.
3. In sugarcane, pre-emergence application of tafazine 50W (Simazine) at 4.5 kg/ha + tafacide 80 (2,4-D sodium 80%) at 1.75 kg/ha is very effective.
4. In groundnut, pre-emergence application of alachlor 1.5 kg/ha

is effective and economical.

3. *Achyranthus aspera*

Prickly chaff flower .... Amaranthaceae
Chirchiri (Hindi)

Achyranthus is an annual coarse weed commonly found in pastures, wastelands and in cultivated maize and Lucerne fields. It becomes troublesome at maturity because of spinous bracteoles and pointed perianth which adhere to the clothes and run into the body of man and animals.

It has a deep tap root; erect slightly branching stem, 50 to 100 cm tall; leaves are opposite, elliptic obovate, 6 to 10 cm long; inflorescence rigid, spikes are rough, long with bracts, having two thorns pointed upward. Flowers small, pale green, fruit is a capsule; seeds are spiny. It is reproduced by seeds.

Control measures

1. In sorghum pre-emergence application of atrazine and simazine with post-emergence application of 2,4-D can control this weed effectively.
2. Atrazine 2 kg/ha and oxadiazone 1.25 kg/ha can give its good control.
3. 2,4-D (ester) at 100 ppm is found to be lethal for Achayranthus.

Economic importance: It is used an antidote for the poison of snakes and scorpions.

4. *Amaranthus spinosus*

Spiny pigweed .... Amaranthaceae
Katalli chauri (Hindi)

Amaranthus spinosus is a troublesome annual herbaceous broadleaf weed, found in cultivated fields, gardens and everywhere,. It has a tap root; erect, hard, branching stem, 75 to 150 cm tall; leaves are simple, alternate lanceolate, 5 to 10 cm long with two sharp spines in leaf axils; inflorescence is monoecious; flowers many, small and sessile, fruit is a capsule. Seeds are dark brown, small and glossy. It is reproduced by seeds.

Allelopathic effect: leaf and inflorescence extracts of *A. spinosus* either spray or soil application, has shown allelopathic affect on mustard and wheat at both vegetative and reproductive phase.

Control measures

1. Ploughing of the land while fallow while fallow and weeding are the common practice to keep off this weed.
2. In crop areas, 2,4-D or MCPB at 1.5 kg/ha as post-emergence or dichlobenil at 3.0 kg/ha as pre-emergence application can give 90% control of this weed.
3. In non-crop areas, PCP at 2 kg/ha; diuron or monuron at 3 kg/ha as pre- or post-emergence can be used.

Amaranthus spinosus

Economic importance: It is used as medicinal plant; ash of this plant is used as dye; leaves are used as leafy vegetable.

5. *Amaranthus viridis*

Slender pigweed .... Amaranthaceae
Jangli Chawri (Hindi)

A. viridis is a common annual weed, widely distributed in the tropics, in cultivated lands, lawn and gardens. It accumulates high concentration of both nitrogen (3.16%) and Potash (4.5%), reported by Vengris *et al.* (1953).

Amaranthus viridis

It has a deep tap root; erect branched stem, 30 to 60 cm high; leaves simple, alternate, ovate, notched at the tip; petioles 10 to 15 cm long; flowers small, bisexual shortly stalked in terminal panicles; fruit compressed; seeds very small, dark brown smooth and shinning. It is propagated by seed, the seed production capacity is about 2,00,000 per plant per year.

Control measures

1. Clean cultivation at early stage by hoeing etc. can keep off this weed.
2. Pre-sowing application of alachlor at 5.0 kg/ha or diuron 1.5 kg/ha is very effective against it.
3. Post-emergence application of 2,4-D or MCPB at 1.5 kg/ha and dichlobenil 3 kg/ha as pre-emergence can control this weed effectively.

Economic importance: ender tops and leaves are used as leafy vegetable.

6. *Anagalis arvensis*

Scarlet pimpernel .... Primulaceae
Krishna neel (Hindi)

Anagalis is an annual creeper; commonly found in sugarcane, linseed and other field crops.

Leaves are alternate and opposite; flowers solitary, blue in colour; corolla tubular, calyx persistent; stamens inserted the petals; ovules numerous; fruit is a capsule, many seeded; seeds angular. It is propagated by seeds.

Control measures

1. MCPA at 0.5 kg/ha or MCPB at 1.0 kg/ha as early pre-emergence can control this weed effectively.
2. In cereal crops, 2,4-D at 1.0 kg/ha may be sprayed only after full tillering phase to control this weed.

7. *Argemone mexicana*

Mexican prickly poppy .... Papaveraceae
Goatweed (Hindi)

Argemone mexicana is an annual herbacious weed of rabi season, grows abundantly in waste areas, road sides and cultivated fields of wheat, chilies, etc. The plant with yellow juice, prickly lobed leaves and yellow flowers are the characteristics.

Argemone mexicana (Pricklypoppy)

It has a tap root; erect, rigid, bluish green stem, 40 to 80 cm high which is prickly throughout. Leaves are coarse, irregularly toothed, alternate, sessile, deeply-lobed and clasp the stem. Flowers are regular and showy; fruit is prickly capsule, having 3 to 5 compartments which split open by valves at maturity. Seeds are numerous, small round and blackish brown with oily endosperm. Seeds are shed and cover the entire field and thus become a nuisance value. Seeds are poisonous.

Control measures

1. Hot weather cultivation and manual weeding before flowering can keep this weed off.
2. 2,4-D (ester) at 125 ml/100/l water as post-emergence, can control this efficiently.

Economic importance: Its oil is used for skin diseases, for itches, in paint industry; also used as lubricant and illuminate.

8. *Asteracantha longifolia*

Acanthus .... Acanthaceae
Satyanashi (Hindi)

Acanthus is a tall herb, found in moist places and also in rice fields. Stem is erect, pinkish, covered with rough hairs. Leaves narrow, found in dense clusters; long, woody spines are formed at the leaf axils. Flowers are two-lipped purple. Fruit is a capsule and seeds much compressed, covered with brown hygroscopic hairs.

Control measures

1. It can be removed by hand weeding in paddy fields at early stages.
2. Chemical weed control with molinate at 2 kg/ha is effective.

9. *Boerhaavia diffusa*

Spreading spider ling .... Nyctaginaceae
Bis-Khapra (Hindi)

Boerhaavia is a perennial herbaceous weed, commonly found in cultivated fields of sugarcane etc; pastures, roadsides and waste-lands. It spreads prostrate and resembles to Trianthema monogyna.

It has deep, fleshy tap-root which stores large amount of food. Stems are semi-woody, diffusely branched, creeping and succulent, swollen at the nodes. Leaves are distantly arranged at the nodes, opposite, obovate and glabrous, upper surface of leaf is green while the lower side whitish with dots. Flowers are very small, red sessile; fruits are glandular and sticky. Seed light brown and winkled. It is propagated by seed-like fruits.

Control measures

1. Monuron or diuron at 5.0 kg/ha as pre-emergence can check this weed.
2. EPTC at 4 l/ha or molinate 2 l/ha as pre-plant application is very effective.

Economic importance: The root of Boerhaavia is successfully used in diarrhea, dysentery and cholera. It is also diuretic and stomachic.

10. *Calotrpis gigantean*

Giant milk weed .... Asclepiadaceae

Giant milk-weed is a perennial woody shrub, common in open wastelands and roadsides. It has a tap root, stem in erect, woody and branched, 75 to 200 cm tall. It produces milky latex which is the chief characteristics of the plant.

Leaves are quite large, 10 to 15 cm in length, oval shape, opposite, blue-green in colour with white veins, pinnately nerved; stipulses absent. Inflorescence is a cyme with pink to white bisexual flowers with a tinge. Fruit is a large follicle, 10 cm long, seeds compressed, crowned with a coma of long silky hairs. It is reproduced by seeds which possess specific organs that keep them afloat and disseminate through wind.

Control measures

1. In waste places, the cluster of this weed should be cut with a hoe.
2. Amitrole at 1% spray can effectively control this weed.

Economic importance: (1) It can be used as a strong purgative and against fever. (2) Decoction of the root is used in infantile convulsions and delirium during fever. (3) Leaves are used for giving warm fomentations in treating abscesses. (4) Root bark and inspissated juice are tonic and diaphoretic.

11. *Carthamus oxycantha*

Wild safflower .... Compositae
Pohli (Hindi)

Carthamus is a persistent weed of winter grain crops, wheat and barley. It is an annual herb, leaves are alternate, simple, flowers crowded into heads, surrounded by bracts; ovule erect from the base; fruit (achene) sessile, seed without endosperm.

Seeds germinate in large numbers during December, set seeds in March-April and shatter on the ground. It is spiny and hence makes harvesting of wheat or barley difficult.

Control measures

1. Rotation of grain crops with berseem or potato can keep off this weed.
2. 2,4-D (Sodium) at 1.2 kg ai/ha as per or post-emergence can kill this weed completely and this chemical control method is economical as compared to hand weeding.
3. Spraying of 2,4-D at 1.0 kg/ha + 2,4,5-T at 1.0 kg/ha as post-emergence is a good combination and very effective for its control.

12. *Cassia occidentalis*

Coffee sena .... Caesalpiniaceae

Coffee sena is an annual shrub like plant, common in waste grounds, road sides, pastures and in perennial crops. It has a tap root; highly branched, woody, erect stem, 30 to 75 cm high; leaves pinnately compound; leaflets have short stalk, lanceolate, opposite and glabrous. Inflorescence is made up of terminal and axillary racemes with yellow flowers. Fruit is dark-brown, 10 to 15 cm long. Seeds are obvate and brown. It is reproduced by seeds.

Control measures

1. Pulling of individual plant in pastures before flowering and clean cultivation in cultivated field, can deep off this weed.
2. Spraying of 2,4,5-T (ester) at 1.5 kg/ha or mixed with 2,4-D (ester) 3:1 ratio with Kerosene or diesel oil can control this weed effectively.
3. 2,4-D sodium (80%) at 10,000 ppm is very effective.

Economic importance: Leaves roots and seeds of coffee sena are used as purgative; these are useful in cough and whooping cough. This plant is also used against skin diseases.

13. *Cassia tora*

Sickle pod .... Caesalpiniaceae
Chakor Chakunda (Hindi)

Cassia tora is an annual plant, found in roadsides, pastures and cultivated fields. It has a long tap root; Stem woody, glabrous, branched, 40 to 100 cm tall; leaves pinnately compound with paired leaflets. Flowers are maxillary, yellow in colour. Seeds are shiny, brown. It is reproduced by seeds.

Control measures

1. Pulling out individual plants should be done on or before flowering.
2. Spraying of 2,4-D (ester) mixed with 2,4,5-T (ester) in 1:3 ratio (1.0 kg/ha each) is quite effective.

Economic importance: It has a medicinal value, constitute a remedy in ringworm and itches; also used for dye and tannin. Young leaves and pods are eaten.

14. *Celosia argentea*

White cock's comb .... Amaranthaceae
Safed murga-ka-phool (Hindi)

Celosia is an annual herbaceous plant, found abundant in cultivated fields, particularly in upland paddy, sorghum and maize.

Celosia argentea

It has a tap root; stem stout, branching, fleshy, erect 50 to 100 cm tall; leaves linear, sessile, alternate; inflorescence is a dense spike. Flowers are pink when young, becoming white later; seeds black and shiny. It is propagated by seeds. It completes its life cycle in 10 to 12 weeks.

Control measures

1. Celosia is sensitive to shading and hence competitive crops can

be grown closely to reduce its spread. Maize is found competitive with this weeds.

2. Since it flowers from 4th week, weeding and clean cultivation is required before 4th week, with rotary weeder or handhoe.
3. 2,4-D at 1.0 kg/ha as post-emergence or simazine 2.24 kg/ha as pere-emergence, can be applied for its control.

Economic importance: (1) Seed is used in diarrhea, blood disease and mouth sores. It is also used for clearing the vision and for eye diseases.

15. *Chenopodium album*

Lambsquarters .... Chenopodiaceae
Bathua (Hindi)

Lambsquarters is an annual weed, found abundantly in all rabi season crops like wheat, pea, potato etc; as a companion of this weed. It accumulates very high concentration of potash (4.34%), nitrogen (2.59%), calcium (1.46%) and also phosphorus and magnesium from the soil, as analyzed by Vengris *et al.* (1953).

It has a tap root; erect stem, 30 to 90 cm high; leaves alternate oblong and lanceolate, petioles slender and long; flowers perfect, having 3 to 5 sepals; petals absent. Seeds black and glossy. It is propagated by seed; its produces 50,000 to 70,000 seeds per plant year.

Control measures

1. Weeding at early stage can done by mechanical means.
2. Pre-emergence application of simazine or atrazine at 2 kg/ha nitrogen or amiben at 3 kg/ha is effective to control pigweed.
3. Post-emergence application of 2,4-D or 2,4-D B or MCPB or 2,4,DES at 2 kg/ha can be applied to kill this weed effectively.

Economic importance: Tender tops and green leaves are used as sag (vegetable); it is rich in potash. Oil is extracted from its seeds.

16. *Cirsium arvense*

Canada thistle .... Compositae
Kateli (Hindi)

Canada thistle was introduced from Canada to India by French for feeding swine. It is an aggressive and pernicious, prickly winter perennial, very persistent and noxious which create special problems for rabi crops like wheat and potato. It interferes in crop culture and also crop harvest.

It is deep rooted, penetrating more than 1 m into the soil; shoot grow about 1 m tall; leaves are opposite, crinkled and spiny; stipules absent; flowers crowded into heads, surrounded by series of free bracts. It reproduces mostly by extensive shallow horizontal root system. Roots remain dormant in soil for many years and hence very difficult to eradicate, particularly in heavily infested areas. It becomes poisonous under undue environmental stress.

Control measures

1. At full tillering stage of wheat, 2,4-D (amine) at 1.5 kg/ha; in Potato, paraquat spray 2.0 kg/ha and in peas, MCPB at 2.0 kg/ha is useful.
2. A combination of MCPB (0.25 kg ai/ha) + MCPA (0.25 kg ai/ha at 3:1 ratio can control this weed effectively.
3. Dicamba at 0.5 kg ai/ha or 2,4-D (0.5 kg) + Mecoprop (1.5 kg/ha) can give excellent control.
4. In non-crop areas, picloram 2 kg/ha, dicamba 4 kg/ha or 2,4-D plus 2,4-5 T at 2:1 ratio is very effective.
5. For biological control, flea beetle is found to be a promising bio-agent.

Cirsium arvense (Canada thistle)

17. *Cleome viscos*

Spider flower .... Capparidaceae
Hurhuria (Hindi)

Cleome is an annual herbaceous plant, found in roadsides and cultivated fields of ragi, sorghum etc. It has a tap root; erect stem, 60 to 150 cm tall. Leaves are simple, palmate and alternate; inflorescence is a terminal receme; flowers bisexual, fruit is a capsule with a long peduncle and cylindrical; seeds are angular, and black. It is reproduced by seeds.

Cleome viscos (Spider flower)

Control measures

1. Hot weather cultivation may keep off this weed.
2. Atrazine at 1.0 kg/ha should be applied before emergence of sorghum or post-emergence application of 2,4-D at 0.75 kg ai/ha may be useful.
3. Chloroxuron 3 kg/ha are found to be selective for ragi.
4. Other weedicides like dicamba 2.0 kg diuron 2.0 kg, alachlor 2.0 kg/ha etc. are also effective.

Economic importance: It is used as leafy vegetable. Leaves are also used in external applications for wounds and ulcers. Seeds can act as stimulant.

18. *Convolvulus arvensis*

Field bindweed .... Convolvulaceae
Hiran-khuri (Hindi)

Plate 4: (1) *Cleome* spp. (Spider flower), (2) *Cucumis melo* wild cucumber

Bind weed is a native of Europe. It is vigorous twinning or trailing, deep-rooted perennial, obnoxious herb. It produces long weak stems twine round other vegetation and pull them down because of its early heavy shoot growth. It is troublesome in wheat, barley, maize, potato, sugarcane and other crops.

It possesses a strong root system, 2 to 3 m deep or even more and is an obligate weed and hence difficult to eliminate. Leaves are obvate-cordate, flowers pink with white streaked petals, funnel shaped corolla; fruits are capsule 2 celled with 2-4 seeds; propagated through seeds also by vegetative growth of the rhizome. It produces plenty of seeds which are capable of germination in about 10 day's time. Rhizomes are spread about 4 metres in a year. Seeds are resistant to mechanical injury.

Convolvulus arvensis (Field bindweed)

Control measures

1. It is very difficult to control this weed because of deep-root system and also under-ground stems break and spread over the field due to careless digging.
2. Application of 2,4-D + dicamba at 0.5 kg ai/ha each or picloram at 0.25 kg/ha at bud stage, can eradicate this weed.
3. Dinoseb 3 kg/ha as early pre-emergence application at tender stage can control this weed successfully.

Economic importance: Roots and rhizomes of field bind weed are used as purgative.

19. *Corchorus acutangulus*

Podrush .... Tiliazeae
Jangli jute (Hindi)

Jangli jute is an annual rainy season weed, found common in roadsides, pastures and also in cultivated fields. It has a strong deep root system; stem woody branched, erect, 50 to 120 cm tall branches spreading; leaves simple, alternate, glabrous with serrated margin and oblong; flowers yellow, borne in axils; fruit bilocular, capsule, 6-angled beaked and cylindrical. Seeds are brownish or blackish. It is propagated only by seeds.

Control measures

1. Clean cultivation by ploughing, mowing etc. can control this weed.
2. While weed seedlings are at 2-3 leaf stage, 2,4-D or 2,4,5-T or MCPA at 2.0 kg/ha can be applied as post-emergence.
3. Diuron 0.75 kg/ha as pre-emergence coupled with MCPB at 2.5 kg/ha can control this weed effectively in cotton.

20. *Cucumis trigonus*

Wild cucumber .... Cucurbitaceae
Gomuk (Hindi)

Cucumis is a herb with watery juice found in fields of maize, sugarcane, pigeonpea etc. often climbing by tendrils, spirally coiled; stems prostrate; flowers yellow, monoecious, calyx tubular, corolla poly-petalous; ovary inferior, ovules numerous and seeds also numerous. Often flattened, without endosperm. It is propagated by seeds.

Control measures

1. Clean cultivation can keep off this weed.
2. Atrazine 2 kg/ha as pre-emergence application is useful for its control.

Economic importance: Fruit is eaten and also used as vegetable.

21. *Datura stramonium*

Thornapple, Kumsonweed ....Solanaceae

Datura is a perennial herbaceous bushy weed, found in fertile fields, pastures and along road sides. It has a thick, shallow, branched tap-root; stem erect 1 to 2 meters tall, branching, glabrous and hollow; leaves ovate with wavy toothed margins, alternate, smooth, long petiole, which are strongly scented. Flowers are large, white; corolla with short stalks. Fruit is a egg-shaped capsule, spiny, having 4 capsule which split open at maturity; seeds black, wrinkled and kidney shaped. It is reproduced by seed, seed production is very high 12,000 to 14,000 seeds per plant of which 5% germinate immediately. Fruits, seeds and seedlings are highly toxic to animal and man.

Control measures

1. Clean cultivation and uprooting at early stage can keep off this weed.
2. 2,4-D (ester) at 3 kg/ha as post-emergence or atrazine 4.5 kg/ha as pre-emergence can control this weed effectively.
3. Pre-emergence application of dinitramine 0.5 kg/ha can be used to control this weed in wheat.

22. *Daucus carota*

Wild carrot .... Umbelliferae

Wild carrot is an annual herb with watery juice. Leaves are alternate; stipules absent. It forms storage organs underground elongated and bi-valved. Seeds are without endosperm. It is mostly found in the field of onion and garlic. The presence of this weed causes reduction in crop stand, number and size of bulbs and thereby the yield of onion or garlic by 40% to 60%.

Control measures

Post-emergence application of oxadiazon 1.0 to 1.5 kg/ha is most effective to control wild carrot. The herbicidal weed control is economical than normal practice of hand-weeding.

23. *Digera arvensis*

Digera .... Amaranthaceae
Kari – gandhari (Hindi)

Digera is an annual, slender weed which is mostly found with kharif season crops all over the country. Leaves are glabrous, ovate, rounded at the apex, margin tinged with red; petioles slender bug; flowers sessile, bracts boat shaped. It is propagated by seed.

Control measures

1. Hoeing and weeding can reduce its spread.
2. Pre-emergence application of picloram 2 kg/ha can effectively control this weed.

Economic importance: It is used as leafy vegetable.

Digera arvensis

24. *Eupatorium odoratum*

Spurge .... Compositae
Communist weed (local name)

Eupatorium is a tropical, perennial weed of Mexican origin, grows in abundance in plantation crops like tea, young rubber, cashew, oil palm, coconut and forest areas; also in road sides and waste lands. It is a poisonous, noxious weed, causing 'trembles' in cattle and 'milk stickiness' in man and hence of national concern.

It is a herbaceous plant with tap root and hardy erect stem; leaves are simple, alternate, serrate at the margin; flowers crowded into heads, surrounded by bracts. Fruit sessile and seed with out endosperm. It is multiplied at its surroundings easily through seeds and hence difficult to eradicate them in the infested areas.

Control measures

1. It can be controlled by complete uprooting in early stage in a newly introduced areas; employing voluntary squads.
2. The effective herbicides against Eupatorium are: MSMA at 3.5 l/ha, a mixture of 2,4-D (1.5 kg/ha) + 2,4,5-T (0.75 kg/ha) i.e. 2:1 ratio, metribuizne 6 kg/ha or glypohsate 6 kg/ha.
3. A mixture of picloram 2 kg + 2,4-D (amino) at 1.5 kg/ha can kill this weed within 6 months and no new weed seedlings will appear.

25. *Euphorbia hirta*

Garden spurge .... Euphorbiaceae
Bari-dudhi (Hindi)

Euphorbia is a common annual herbaceous weed with milky juice, grows abundantly in all types of soil in road sides and cultivated fields with special affinity with wheat, bajra, pigeonpea, green gram, cotton, groundnut, sugarcane, turmeric.

It has a tap root; stem reddish, branched, pubescent and semi-erect. Leaves lanceolate, opposite, elliptic oblong with acute tip; internodes short thus making leaves more closer. Inflorescence consists of axillary clusters of composite flowers, monoecious; male flowers are composed at naked pedicelled stamens; capsule hairy; seed oblong and reddish. It is propagated by seed.

Control measures

1. Hot weather cultivation can keep off this weed.
2. Alachlor 1.5 kg/ha as pre-emergence is found to be effective in cotton and turmeric.
3. Asulox 1.5 kg + 2,4-D ester 1.5 kg/ha in sugarcane or Prometryne 2 kg/ha in groundnut can give a good control of this weed.

Euphorbia hirta (Spurge)

Economic importance: Root is given to induce vomiting and the plant is given to nursing mother.

26. *Gynandropsis pentaphylla*

Wild spider flower .... Capparidaceae
Hurhur (Hindi)

Gynandropis is an annual herbaceous strong smelling weed of waste places. The plant is distinguishable by its strong disagreeable odour and sticky glandular hairs.

It has a deep tap root; stem freely branched, erect 50 to 100 cm tall, hairy and sessile bracts. Leaves are alternate, palmately compound, long petioled with 5 leaflets. Inflorescence recemose, terminal with white flowers; fruit is a capsule, 6 to 10 cm long; seeds small, kidney-shaped and dark brown. It is propagated by seeds.

Control measures

1. Two to three hoeings can be given to remove this weed.
2. Prometryne 0.6 kg/ha as pre-emergence application plus one hand weeding after a month can be given.
3. Tafazine 50 W at 6.5 kg/ha is suitable for controlling this weed in sugarcane.
4. Dinitramine 1 to 2 kg/ha is proved very effective in upland paddy.

27. *Indigofera glandulosa*

Indigo ....Caesalpiniodea
Neel (Hindi)

Indigo is an annual herb, common on river banks, and in the drier regions. Stems are slender, prostrate and pubescent; leaves linear oblong; flowers small, bright red, calyx tube as long as the pods, silky. It is propagated by seed.

Control measures

1. Weeding and hoeing are given to control this weed.
2. Atrazine 1.5 kg/ha as pre-emergence or Silvex 5 kg/ha as post-emergence causes killing effect of *Indigofera* sp.

28. *Ipomoea purpurea*

Morning glory ....Convolvulaceae
Karni (Hindi)

Ipomoea reptans is a vigorous evergreen herb, but in extreme drought, it undergoes long dormancy period. It is a predominant weed in semi-arid regions, mostly in non-cropped areas. In crop fields, it is troublesome in wheat, barley, oats, sugarcane and potatoes. It binds the crop plants and makes their harvest extremely difficult and cumbersome. Seeds have a long dormancy period. It is propagated by seeds and root stalks.

Control measures

1. Summer tillage and growing fodder crops in rotation are good cultural methods to control this deep-rooted troublesome weed.
2. 2,4-D (ester) at 1% concentration is very effective for its control.
3. In potatoes, post-emergence application of paraquat (1.5 l/ha) is found to be effective.
4. 2,4-D at 0.05 kg/ha + dicamba 0.5 kg/ha or Picloram 0.25 kg/ha at early stage, can control this weed effectively.

29. *Lantana camara*

White sage, brush .... Verbenaceae
Phul-bakri, putus (Hindi)

Lantana is a perennial semi-woody evergreen shrub collectively called 'brush'. It was introduced from central America as an ornamental plant and has now become a weed in waste land and road sides.

Lantana camara (white sage)

It has a tap root; stem woody, much branched, rough and thorny; leaves pubescent opposite, ovate and toothed; inflorescence is composed by racenus borne on a long peduncle that originates on the leaf axils, bears multicolour flowers (red, yellow or orange), clustered in compact heads. Fruit is a spherical drupe, dark-purple splits into two seeds. It is reproduced by seeds.

Plant has a fragrant odour and contains alkaloids which cause high mortality among livestock. It produces a toxic constituent 'Lantadenes' which causes severe damage of the liver and produces symptoms of jaundice.

Control measures

1. Mechanical control may check its spread.
2. A combination of 2,4-D 1 kg + dalapon at 5 kg al/ha is very effective for its control.
3. Smothering cut plants with spontox (mixture of 2,4,5-T) 25% solution can stop its sprouting for a few months.
4. Biological control has been successful in Hawali. Certain insects bio-agents like *Hypena jusatis* is found to be an excellent feeder of lantana leaves; while the moth, *Crocidosem lantana* destroys its flowers and seeds.

Lathyrus aphaca

30. *Lathyrus aphaca*

Lathyrus .... Papilionaceae
Matari, pipura, (Hindi)

Lathyrus is a herb, found in rabi season in cultivated fields, associated with wheat, barley, gram etc. Leaves are reduced to tendrils, at the base of which large foliaceous stipules perform the function of leaflets. Flowers yellows which produce pods. Seeds are black, smooth and compressed. It is reproduced by seeds.

Control measures

1. It can be removed by mechanical means at its first appearance in the fields.
2. Isoproturon (50 WP) at 1.0 kg/ha is most effective in controlling this weed.

31. *Leucas aspera*

Leucas .... Labiatae
Guma (Hindi)

Leucas is commonly found in wasteland and also cultivated fields of wheat, maize, pigeonpea etc.

Plant is profusely branched, growing to a height of 45 cm; stem is covered with coarse hairs; leaves are narrow and opposite; flowers white, crowded in dense terminal clusters. It is propagated by seed.

Control measures

1. Hot weather cultivation can keep off this weed.
2. Asulox 40 + 2,4-D ester is quite effective and comparable to fortnightly hand-weeding.

Economic importance: Juice of this plant is used in snake bites, cut injury and ring worm.

Melilotus alba

32. *Melilotus alba* Sym. *vulgaris*

White sweet clover .... Papilionaceae

Melilotus alba is an annual rabi season leguminous weed, associated with Lucerne crop. It produces white flowers. It is propagated by seeds.

Control measures

1. Clear cultivation is the common practice.
2. Pre-emergence application of linuron 1.0 kg/ha, 2 weeks after emergence, can control this weed effectively.

   Economic importance: It is an anti-blood coagulant.

33. *Meliolotus indica* Syn. *parviflora*

Senji .... Papilionaceae

Senji is an annual rabi season leguminous weed mostly found in waste lands. This is a spreading type of weed. It produces yellow flowers. It is propagated by seed.

Meliolotus indica

Control measures

1. Clean cultivation would keep off the weed.
2. EPTC at 3.0 kg/ha as pre-planting maybe incorporated in the soil, followed by post-emergence application of nitrogen 2 kg/ha at 3 to 4 leaf stage, to control this weed effectively.

34. *Mimosa pudica*

Sensitive plant, touch-me-not ....Mimosaceae
Lajbanti (Hindi)

Mimosa pudica is a perennial woody shrub, found in waste lands, pastures, cultivated drylands and plantation crops. Leaves fold up rapidly when touched or at sunset and hence it I scalled 'touch-me-not'.

Stem is spiny and pubescent; somewhat trailing habit and produces roots at each node; thus difficult to eradicate. Leaves bi-pinnate with 4 primary leaves, originating from tip of each petiole; axillary, consists of a pedunculate flower head; flower pink, with prominent stamens. Pods occur in clusters, comes to a sharp point at the tip. It is reproduced by seed.

Control measures

1. Post-emergence application of propanil 4 kg/ha in combination with Oxadiazon 3 kg/ha can control *Mimosa* very effectively in dryland rice.
2. Alachlor 1.5 kg/ha is used as pre-emergence herbicide for its control in cotton.

35. *Mollugo verticillata*

Carpet weed ....Molluginaceae

Mollugo is a herbaceous weed with fleshy leaves. Leaves are opposite; stipules small; flowers small, inconspicuous, bisexual; sepals free; petals small; ovary several locular; fruit is surrounded by persistent calyx, seeds with endosperm.

Control measures

1. Pre-emergence spray of bromacil 3 kg/ha can keep the Pineapple field weed free.
2. Dinitramine (USB 3584) is proved to be very effective in controlling this weed in upland rice.

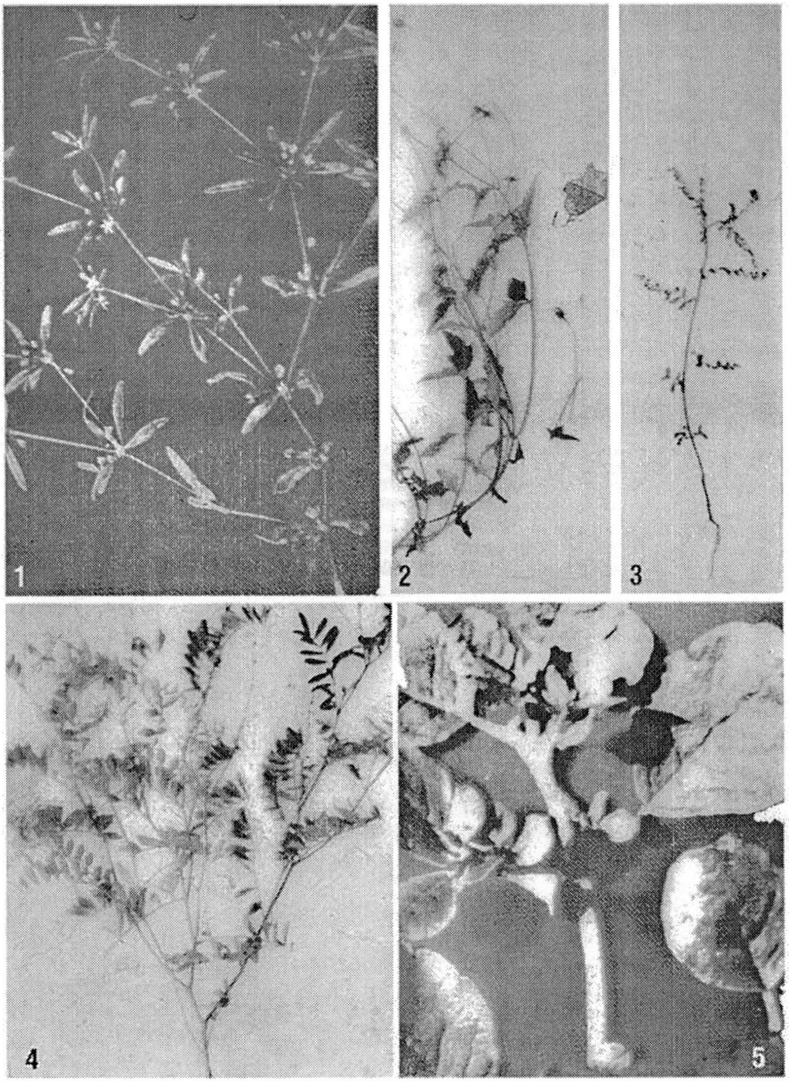

Plate 5: (1) *Mollogo verticillata* (Carpet weed), (2) *Tridex procecmbens* (Tridex), (3) *Phyllanthus niruri* (Gripe weed) (4) *Tephrosia purpurea* (Wild indiga), (5) *Trianthema portulacas trun* (Horse purisilam)

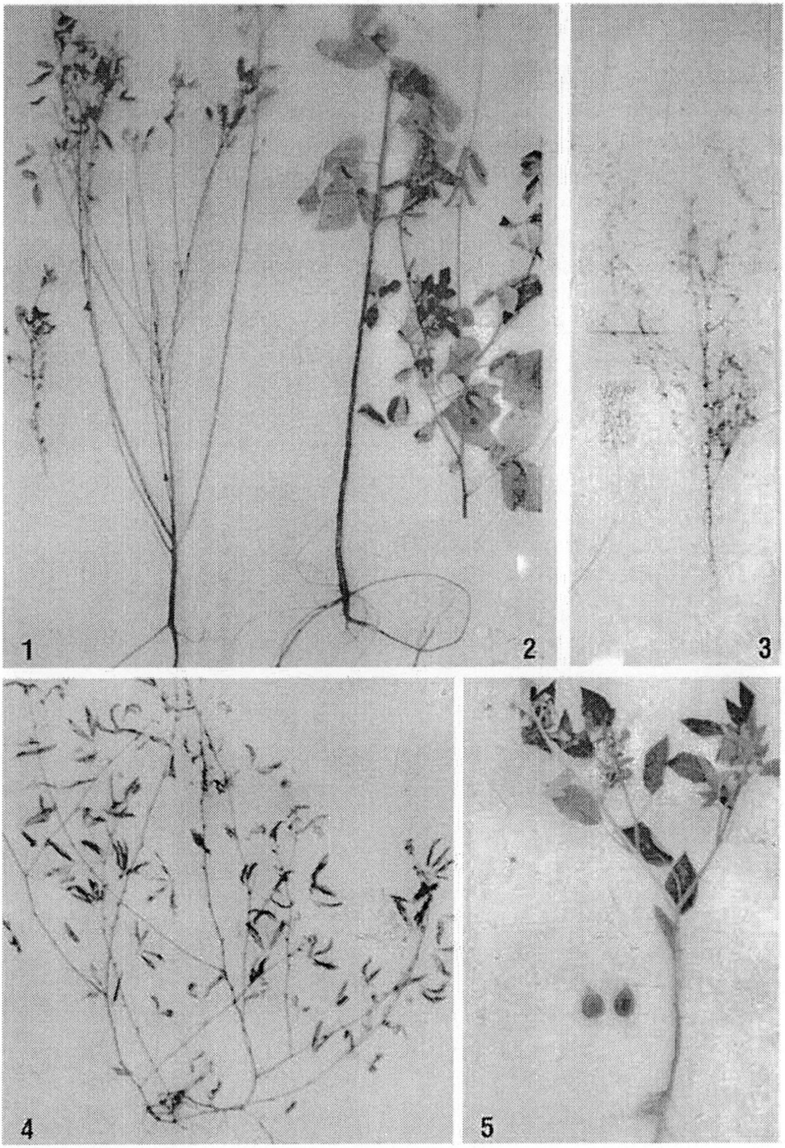

Plate 6: (1) *Meliotus indica* (Yellow sweet clover), (2) *Meliotus alba* (White sweet clover), (3) *Mollugo verticillata* (Carpet), (4) *Mimosa pudica*, (5) *Physalis angnlata* (Ground cherry, hogweed)

36. *Opuntia dillenii*

Prickly pear .... Cactaceae
Nagphani (Hindi)

Opuntia is a perennial xerophytic succulent and facultative weed which grows mostly in wild communities. It has large fleshy flat pads, equipped with prickle. It has inconspicuous scaly, much reduced leaves which usually drop off. Each plant produced several storied pods and each pod can produce a new plant. Flowers open in the morning and close in the evening; fruit is a berry, many seeded which are immersed in the pulp. Berries are eaten by birds who disseminate the seeds. It is also propagated by means of cuttings.

Control measures

1. Mechanical control measures should be adopted to eradicate it from a localized area.
2. 2,4,5-T and Picloram at 2 kg/ha each can be sprayed for its control.
3. In severely infested areas, biological control has been found very effective in Australia and subsequently in India. The bioagents, *Cactoblastis cactorum* in Australia and *Dactylopius tomentosus*, a cochineal scale insect in India, are found to be the best to control prickly pear.

Economic importance: It is used as hedge for fencing and also used as medicinal plant.

37. *Oxalis acetosella*

Wood sorrel, lady's sorrel ....Oxalidaceae
Amrauna (Hindi)

Oxalis is a perennial herbacious weed, common at roadsides, waste places and also found in cultivated fields in sandy soils. It is particularly noxious in tea, coffee and rubber.

Stem is diffused with procumbent branches and covered with loose curly hairs. Leaves alternate, palmately compound, trifoliate leaflets, sessile, spirally coiled, usually folded at night. Flowers yellow, perfect and regular. It flowers and fruits year round. Fruit is a capsule. Seeds are flattered with an elastic testa. It is propagated by seeds and stocks of roots and stem.

Control measures

1. Clean cultivation can control this weed.
2. Application of 2,4-D (butyl ester) at 5 kg/ha is very effective.
3. Diuron 2 kg/ha or nitrofen 5 kg/ha as post-emergence application is also effective.

38. *Parthenium hysterophorus*

Ragweed .... Compositae
Congress gras, Gajar Gavet, Carrot grass, Pandhari Phule, Chatak chandani (local name)

Parthenium is a pernicious exotic American weed, recognized in India in 1956 which has assumed importance in recent years as a noxious weed, due to profuse growth, fast multiplication, raid spread, agricultural and health hazards. It completes fast with natural useful vegetation and suppress it. it has dominant characters such as prolific seed bearing, rapid dissemination and effective spread allover. *Parthenium* is a dominant type, when once it is established, it does not allow other plants to grow, due to toxic secretion from roots. Thus it is a highly troublesome weed which has nuisance value.

Plant grows to a height of 1 m; stem is profusely branched, hairy, longitudinally grooved; leaves are alternate, simple, pimately compound, tips dissected, margin entire, pubescent on both sides; venation prominent on lower surface. Inflorescence is terminal, peduncled and hairy; each bract sub-tending a female flower with 2 male flowers one in each side. Fruit is of ellipsoid shape, obovate, dark brittle.

Allelopathic effect: Plant is toxic in all its including root system. Leaves are causing allergic dermatitis, pollen causes persistent running from the nose and cold, reddening of eyes with swelling of yields.

Beneficial use: In recent years, chopped Parthinium weeds along with cow dung and old slurry in the proportion of 1:4:1 respectively, are mixed to generate about 1440 cc of bio-gas per day, over a period of 12 weeks.

Control measures

1. It needs integrated approach involving mechanical, chemical, biological and social measures. All concerted efforts are needed towards its eradication.
2. It can be controlled in early stages by complete uprooting by

mechanical means.

3. Mixture of 2,4-D at 2 kg/ha + paraquat 2 l/ha spray is very effective because of their translocative and contact action, respectively.
4. In maize or sorghum field, atrzine 1.5 kg/ha or chlorobromuron 1 kg/ha or monuron 0.75 kg/ha as pre-emergence application is effective. Atrazine prevents seedling emergence of Parthenium in cereals for at least 6 months.
5. In cowpea, guar, groundnut and orchards, DSMA 2 kg/ha as post emergence; while in vegetable and banana fields, alalchlor 2 kg/ha; and in case root-crops, nitrofen 2 kg/ha is very effective.
6. In non-cropped areas high dose of MSMA (4 l/ha) in 400 L water has been found to control it effectively.
7. Diquat at 2 kg/ha cause 100% mortality of this weed within 15 days.
8. Per-emergence application of Oxadiazon 1.25 kg, simazine 2 kg and dalapon 6.0 kg/ha are also effective.
9. Sinbar 0.15 kg plus 2,4-D (amine) l/ha and Reglone 2.5 l/ha as post-emergence sprays are most effective.

Phyllanthus niruri (Jangli amli)

39. *Phyllanthus niruri*

Gripeweed .... Euphorbiaceae
Hazardana (Hindi)

Phyllanthus is an annual monocious weed with milky juice, common in roadsides and cultivated fields. It is grown vigorously in moist soils in paddy, sorghum, cotton, groundnut, soybean etc.

It has a tap root; stem is glabrous; erect, 25 to 80 cm tall; leaves pinnately compound leaflets are opposite, oblong and lanceolate; flowers white imperfect, borne on leaf axils. Fruit is a capsule. It produces orange triangular seeds.

Control measures

1. Clean cultivation can keep off this weed.
2. Dinitramine (USB) (3584) at 2 kg/ha can be applied 3 days after sowing of paddy.
3. Dinitramine 0.8 kg/ha can keep off the weed for about 6 weeks in cotton, groundnut, soybean and chilies.
4. 2,4-D at 0.75 kg ai/ha is also very useful in kill this weed in sorghum.

Economic importance: It has some medicinal value. Plant is used as de-obstruent, diureticastringent and cooling. Juice of leaves is a remedy for bleeding of gums. It is administered in jaundice, dropsy and genitor-urinary affections.

40. *Physalis minima*

Hogweed, groundeherry .... Solanaceae
Wild. Goose berry

Physalis is an annual herb, common in roadsides, river banks and cultivated fields of sorghum, sugarcane etc.

It has a taproot; stem erect, 75 to 100 cm tall, branched, angular and hollow; leaves alternate, smooth ovate with irregular margins; flowers yellow and characterized by 5 partially united petals which later enclose the fruit. Fruit is a berry; seeds small, flattered and kidney shaped seeds possess specific organ called 'balloon'; a modified papery calyx, than encloses the fruit loosely with entrapped air. It is reproduced by seed.

Control measures

1. Simazine at 2.0 kg as pre-planting or 1.5 kg/ha as pre-emergence and atrazine at 1.0 kg/ha before crop emergence is a

potent herbicide to kill this weed in sorghum.
2. Diuron 2 kg ai/ha as pre-emergence is also a good killer of this weed in sugarcane.
3. 2,4-D at 0.75 kg/ha as post-emergence is found to be useful.

41. *Pluchea lanceolata*

Arrow wood .... Compositae
Baisuri (Hindi)

Pluchea is a perennial semi-woody thorny noxious weed, occurring all over India and hence of national concern. It is commonly found in wheat field.

It is deep rooted, throws out green shoots after winter season and remains green in summer. It flowers and sets seeds and then shoots dry up, but roots undergo dormancy.

Control measures

1. Paraquat at 4.5 l/ha in two sprayings can check its growth.
2. Mixture of 2,4,5-T (0.5 kg/ha) + 2,4-D (1.0 kg/ha) at 1:2 proportion applied in a wetting spray volume is very effective.
3. Spraying of Agroxone 10% (Methoxone) in 960 1 water/ha water can completely destroy this weed in a heavily infested area.

42. *Polygonum plebejum*

Smartweed .... Polygonaceae
Raniphul (Hindi)

Polygonum is an annual prostrate herb, commonly found in waste lands and also in cultivated fields of sugarbeet, pigeonpea and gram. It is observed that both *Polygonum* and *Chenopodium* constitute more than 75% of the total weed population in the infested fields.

Stem is grooved, branched; leaves linear alternate, apex acute and stipules fimbricate; flowers arranged in clusters, small, pink with green mid-rib. Fruits are spindle shaped with abundant seeds. It is propagated by seed.

Control measures

1. Hand weeding is the common practice to control this weed. Mechanical weeding after one month of planting of pulse crops like pigeonpea, keep the field weed free.
2. Pre-emergence application of prometryne 0.25 kg ai/ha,

alachlor 1 kg ai/ha or nitrofen 2.0 kg/ha is very effective.

43. *Portulaca oleracea*

Purslane ....Portulacaceae
Kulfa-sag (Hindi)

Portulaca is an annual prostrate, succulent herb, common in both garden lands and non-cultivated waste grounds. It prefers rich soils growth, accumulates high concentration of potash (4.5%) and nitrogen (2.4%) and thus become a great competitor for crop plants like cotton, groundnut, soybean, chilies etc.

It has a tap-root system; stem short, erect, bears lateral prostrate branches, forming mats on the ground. Leaves are succulent, smooth and fleshy, flowers are small, simple, sessile axillary, bi-sexual forming clusters and supported by a whorl of leaves; fruit is a spherical capsule; seeds flattened, black and minute. It is propagated by seed. It is prolific in seed production, producing as high as 1,50,000 to 2,00,000 seeds by a single plant. Seed viability persists even for 25 to 30 years.

Control measures

1. Dicamba at 1.5 kg/ha as post-emergence; 2,4-D-B (ester) 0.75 kg/ha or trifluralin 0.75 kg/ha as post-planting application can control this weed effectively.
2. Dinitramine at 0.8 kg/ha control this weed upto 6 weeks in cotton, groundnut and soybean.

Economic importance: It is used as very good green vegetable.

44. *Solanum xanthocarpum*

Night shade ....Solanaceae
Kateli (Hindi)

Solanum xanthocarpum is a conspicuous rabi season weed, common in black soils throughout plains.

It has a deep root system upto 1 m deep; branches spread on the ground, having numerous long spines all over the plant which are the distinguished characteristics of this weed. Flowers are showy, purplish blue in colour; fruit is a berry with green blotches. It is propagated by seeds.

Control measures

1. It can be controlled by mechanical weeding at early stage.

2. Picloram plus biomacil combination at 5 l/ha each can will this weed completely within 3 months and keep the field weed free; while 2,4-D (sodium) at 2.5 kg/ha spray can keep off this weed within 8 weeks.

Economic importance: Fruits are boiled in butter and given for cough and toothache.

45. *Sonchus* spp.

Sowthistle .... Compositae

Sonchus pp. is a winter season weed, common in cultivated fields of wheat, barley, linseed etc. It is grown along with Canada thistle and becomes more troublesome in the cultivated crop field.

It is a herbacious plant, leaves opposite, stipules absent, flowers crowded into heads, surrounded by tracts.

Control measures

A combination of MCPA at 0.25 kg/ha plus MCPB at 0.25 kg/ha is most effective.

Sonchus arvensis (Sowthistle)

46. *Spergula arvensis*

Cron sparry .... Caryophyllaceae
Bandhania (Hindi)

Spergula is an annual green herb, found in cultivated fields of sugarbeet, potato, pea, gram etc.

Leaves are opposite, single, entire; flowers bisexual, white;

sepals united into a tube; petals absent; fruit is a dry capsule, opening by valves. Seeds are narrowly winged with endosperm. It is propagated by seed.

Control measures

1. Mechanical weeding is the common practice to remove this weed.
2. Alachlor (lasso) at 3 kg ai/ha as pre-emergence can be used in sugarbeet.
3. In pea, metoxuron 2 kg, metribuzin 0.5 kg or terbutryn 0.5 kg/ha is proved to be very effective.

47. *Stellaria media*

Chick weed .... Caryophyllaceae

Stellaria is an annual herb, fond mostly in wheat fields. Leaves are opposite, connected at the base; flowers bisexual; ovules numerous; fruit is a capsule.

It is well adapted to maturing its seeds with the crop and even uprooted plants.

Control measures

1. Alachlor at 3 kg ai/ha as pre-emrgence application can control this weed effectively.
2. Terbutryn at 0.5 kg/ha is also effective.

48. *Trianthema monogyna* syn. *portulacastrum*

Horse purselane ....Aizoaceae
Patharchatta (Hindi)

Trianthema is an annual rainy season succulent herb, common in waste lands and orchards. Stem is round, prostrate, hairy, branched forked upto 20 to 40 cm long, fleshy; nodes prominent. Leaves are fleshy, opposite petioeld, round to oval, often purple in colour, margins wavy, glabrous; flowers axillary, sessile, often white; fruit is a capsule, hidden in the petiole sheath; seeds dull-black, with hard seed coat, kidney shaped having concentric lines. It is propagated by seed. Farmyard manure is the chief source for its introduction in new areas. It can produce more than 50,000 seeds per plant.

Allelopathic effect: It is considered as a poisonous plant because of its high oxalate content. It is found to be salt exudating in nature.

Control measures

1. Mechanical methods are kharif crop harvest will keep off this weed.
2. Atrazine at 1.5 kg/ha as pre-emergence can destroy this weed in maize. Atrazine with alachlor 1.0 kg/ha is more effective than single application.
3. Dicamba at 1.5 kg/ha as post-emergence application is the best to kill this weed.

49. *Tribulus terrestris*

Puncture vine .... Zygophyllaceae
Gokhru (Hindi)

Puncture vine is an introduced weed from South Europe, now occurs throughout plains, especially in dry sandy sols. It is an annual herb, conspicuous with its prostrate habit and spiny fruits. It has a tap root; stems are densely hairy, branching from the base to form dense mats of slender trailing branches, compound, leaflets oblong and small. Flowers are axillary peduncles. Fruit spherical, has 3 to 5 sharp thorns which can penetrate even shoes and tyres. Seeds arè brown and have to spines. It is reproduced by seeds.

Tribulus terrestris

Control measures

1. 2,4-D at 3.5 kg/ha is effective; however a mixture of 2,4-D (2 kg) + 2,4,5-T (1.0 kg) in 2:1 ratio as post emergence application is effective for its control.

2. Propazine or atrazine at 0.75 kg/ha as early post-emergence is effective.

Economic importance: Dried fruits are used in Ayurvedic medicines; especially in diseases of genitor-urinary system.

Major Dicotyledonus Weeds (Family wise)

Family	Scientific name	Common name	Local name
Acanthaceae	*Asteracantha longifolia* Nees. *Justicia prostrate* Gamble.	Acanthus	Satyanashi, Derua
Aizoaceae	*Trianthema monogyna* L. Syn. *Trianthema portulacastrum* L.	Horse purselane	Patharchatta
Amaranthaceae	*Amaranthus dubius* Mart.	Pigweed	
	A. spinosus L.	Spiny pigweed	Kataili
	A. viridis L.	Pigweed	Chaurai
	Achyranthes aspera L.	White cock's comb.	Safed Murga-ka-phool
	Digera arvensis Forsk.	Amaranthus	Karigandhari
Asclepiadaceae	*Calotropis gigentea*	Giant milkweed	Akona
Cactaceae	*Opuntia dillenii* Haw.	Prickly pear	
Caesalpiniodea	*Cassia occidentalis* L.	Coffee senna	Chakora
	Cassia tora	Sickle pod	Chakor chakunda
	Indigofera glandulosa wild.	Wild indigo	Neel
Capparidaceae	*Cleome viscose* L.	Spider flower	Jungle hurhur
	Gynandropsis pentaphila	Cats	Hurhur
Caryophyllaceae	*Spergula arvensis* L.	Corn sparry	Bandhania
	Stellaria media (L.) vill.	Chickweed	Chenopodiaceae
Chenopodiaceae	*Chenopodium album* L.	Lambs quarters	Bathuna
	Chenopodium murale L.	Neetle leaf, goose foot	
	Salsola kali L.	Russian thistle (Salt wort)	
Compositae	*Acanthosperum hispidum*	Bristly starbur	
	Ambrosia artemistifolia L.	Common ragweed	
	Carthamus oxyacantha	Wild safflower	Pohli
	Cirsium arvense L.	Canada thistle	Kateli
	Eupatorium odoratum	Spurge	Communist weed
	Lagasca mollis	Skeleton weed	
.	*Parthenium hysterophorus* L	Ragweed	Congress grass

Family	Scientific name	Common name	Local name
	Pluchea lanceoalta Oliv	Arrow wood	
	Sonchus arvensis L.	Perennial sowthistle	
	Tridax procumbens L.	Tridax	Tunaki
	Xanthium strumerium B.	Cocklebar	Chhota-gokhra
Convolvulaceae	*Convolvulus arvensis* L.	Bindweed	Hirankhuri
	Ipomoea purpurea	Morning glory	Karni
	Cuscuta reflexa	Dodder	Akash bel
Crucifereae	*Brassica vaper* Dc.	Wild mustard	
	Chorispora tenella	Blue mustard	
	Daucus carota	Wild carrot	
	Raphanus raphanistrum	Wild radish	
Curcurbitaceae	*Coccinia indica*		
	Cucumis trigonus Roxb.	Wild cucumber	
Euphorbiaceae	*Acalypha indica* L.	Copper leaf	Kuppi
	Euphorbia hirta L.	Garden spurge	Bandudhi
	Euphorbia maculate / supine	Spotted spurge	
	Phyllanthus niruri L.	Gripewed	Hazardana
Labiatae	*Leucas aspera*		Guma
Malvaceae	*Hibiscus vitifolius*		
Mimosaceae	*Mimosa pudica* L.	Touch-me-not	Lajjavanti
Molluginaceae	*Mollugo vericillata*	Carpetweed	
Nyctaginaceae	*Boerhaavia diffusa*	Creeping spiderling	
Orobanchaceae	*Orobanche aegyptiaca* Pers.	Broom rape	Hadda, thokra
Oxalidaceae	*Oxalis acetosella*	Wood sorrel	Amrauna
Papaveraceae	*Argemone mexicana*	Mexican prickly poppy	Goat weed
Popilionaceae	*Lathyrus aphaca*	-	Matari
	Medicaga lupelina	Bur clover	
	Melilotus indica	Yellow sweet clover	
	Vicia sp.	Vetch	Ankta
Papilionaceae	*Clitoria ternatea*		
Polygonaceae	*Polygonum plebejum*	Smart weed, knot weed	
Portulacaceae	*Portulaca oleracea* L.	Common perslane	Lunuk

Family	Scientific name	Common name	Local name
Primulaceae	*Anagallis arvensis* L.	Scarlet pimpernel	Krisnaneel
Scrophulariaceae	*Striga asiatica* Lour.	Figwort agiakhar	
	Striga densiflora Benth,	Witchweed	Agiakhar
Solanaceae	*Datura stramonium* L.	Thorn apple	Datura
Solanaceae	*Physalis minima* L.	Hogweed, groundcherry	Banamakoya
Solanaceae	*Solanum xanthocarpum*	Nightshade	Kateli
Solanaceae	*Solanum elacagnifolium*	Silverleaf	
Tiliaceae	*Corchorus acutangulus* Lank.	Pod bush	Ban jute
Verbenaceae	*Lantana camara* L.	White sage	Phulbakri
Zygophyllaceae	*Tribulus terrestris* L.	Puncture vine	Chhotagokhru

Chemical control of Dicots

Weeds	Effective herbicides (kg or l/ha)
Acalypha indica	Stam F-34 (2.5), Basalin (1), EPTC (3)
Acanthospora hispidium	Basalin (1), Tafazin 50 W (4.5), Alachlor (1.5)
Achyranthus aspera	Atrazine (2), Oxadiazine (1.25), 2,4-D acid (100 ppm)
Amaranthus app.	2,4-D (1.5), MCPB (1.5), Dichlobenil (3), PCP (2) Alachlor (5), Diuron (1.5)
Anagalis arvensis	MCPA (0.5), MCPB (1.0), 2-4-D (1.0)
Argemone maxicana	2,4-D (1.5)
Asteracantha longifolia	Molinate (2)
Boerhaavia diffusa	Monuron (5.0), Diuron (.50), EPTC (4)
Calotrpis gigantean	Amitrole (1%)
Carthamus oxycantha	2,4-D Sodium (1.2), 2,4-D (0.75) + 2,4,5-T (0.75) (1:3 ratio)
Cassia spp.	2,4,5-Tester (1.5), 2,4,5-T + 2,4-D (3:1 ratio)
Celosia argentea	2,4-D (1.0), Simazine (2.75)
Chenopodium album	Simazine (2.0), Atrazine (2), Nitrofen (3.0), Amiben (3.0)
Cirsium arvense	2,4-D amine (1.5), MCPB (0.25) + MCPA (0.25) at 3:2 ratio. 2,4-D + 2,4,5-T (2:1) ratio), Dicamba (0.5) 2,4-D (0.5) + Mecoprop (1.5)
Cleome viscose	Atrazine (1.0), 2,4-D (0.75), Chloroxuron (3) Chloramben (3)
Convolvulus	2,4-D (0.5) + Dicamba (0.5) Dinoseb (3) (2,4-D (2), 2,4,5-T (2),

Weeds	**Effective herbicides (kg or l/ha)**
arvensis	MCPA (2) Diuron (0.75)
Cucumis trigonus	Atrazine (2)
Cuscuta sp.	Dichlobenil (2), chlorpropham (2)
Datura stramonium	2,4-D-ester (3), Atrazine (4.5), Dinitramine (0.5)
Digera arvensis	Picloram (2)
Eupatorium spp.	MSMA (3.5) 2,4-D + 2,4,5-T (2:1 ratio) Glyphosate (6), Picloram (2) + 2,4-D amine (1.5)
Gynandropis pentaphylla	Prometryme (0.5)
Ipomoea purpurea	2,4-D ester (1% conc), 2,4-D (0.5) + Dicamba (0.5), Picloram (0.25)
Lantana camara	2,4-D (1.0) + Dalapon (5), Spontox 25% solu.
Loranthus sp.	Paraquat (3.5) + 2,4-D (3.5)
Melilotus spp.	Linuron (1.0), EPTC (3) Nitrofen (2)
Opuntia dillenii	2,4,50T (2), Picloram (2)
Orobanche sp.	Glyphosate (0.5), Dalapon (6)
Oxalis latifolia	2,4-D (5), Diuron (2), Nitrofen (5)
Parthenium hysterophorus	2,4-D (2) + Paraquat (2), Atrazine (1.5), Alachlor (2)
Phyllanthus niruri	Dinitramine (0.8), 2,4-D (0.75)
Physalis minima	Simazine (2), Atrazine (1.0), Diuron (2), 2,4-D (0.75)
Pluchea lanceolata	Paraquat (4.5), 2,4,5-T + 2,4-D (1:2 ratio) Methoxone 10% solu.
Polygonum plebejum	Prometryne (0.25), Alachlor (1.0), Nitrofen (2.0)
Portulaca oleracea	Dicamba (1.5), 2,4-D-B (0.75), Trifluraline (0.75) Dinitramine (0.8)
Solanum xanthocarpum	Picloram (5) + Bromacil (5), 2,4-D sodium (2.5) Tordon (6)
Sonchus spp.	MCPA (0.25) + MCPB (0.25) at 2:3 ratio.
Spergula arvensis	Alachlor (3), Metoxuron (2)m Metribuzin (0.5)
Stellaria media	Alachlor (3), Terbutryn (0.5)
Striga asiatica	2,4-D amine (1.2), Dicamba (2), Fenac (2)
Trianthema spp.	Atrazine (1.5), Atrazine (1.0) + Alachlor (1.0), Dicamba (1.5)
Tribulus terrestris	2,4-D (3.5), 2,4,D + 2,4,5-T (2:1 ratio) Propazine (0.75), Atrazine (0.75)

CHAPTER 4

Parasitic Weeds and their Control

Parasitism is a biological mechanism to gain something for nothing. Parasitic weeds grow among the tissues of host plant, absorb mineral nutrients directly from the host crop and destroy them in the long run. The most serious parasitic weeds are –

(a) *Orobanche aegyptiaca* Pers. (broom rape) and *Strige luta* Lour. (fig. wort) which are host-specific, angiospermous root parasites; while (b) *Cuscuta* spp. (dodder) and *Loranthus* spp. Are stem parasites.

1. *Cuscuta* spp.

Dodder ....Convolvulaceae
Amarlati, Akashbel (Hindi)

There are two main *Cuscuta* species (*C. campestris* and *C. trifolii*). *Cuscuta* spp. Is a parasitic leafless twining herb with golden yellow, shining wiry stems, found on trees, cultivated crops and hedges.

It develops roots at early stage of development and once attached to the host, the roots disintegrated. Stem is thread-like, cylindrical and much branched. It forms dense masses which smother the host plant. It also develops haustoria which penetrate the host plant, extract water, receive nutrients and certain stimuli. It has rudimentary scale leaves and small cluster of white or yellow flowers. Fruit is a round capsule; seeds are round and rough. It is reproduced by seeds as well as fragments of thread like stems, carried by man and animals unknowingly. It is host specific.

Seeds germinate readily, put out haustoria on a true host and multiplies by its thread-like stems rapidly.

Control measures

1. *Cuscuta* can be pulled out and buried. Crop rotation is essential.

2. Paraquat 1% spray with a high-volume sprayer will kill dodder.
3. Dichlobenil 2 kg/ha or chloropropham at 2 kg/ha applied in granular form in irrigated areas, can control dodder for 6 weeks or more.

Economic importance: Threadlike stems are used for headache and gout. Powder of the dried plants is applied on fresh wounds, this powder is also used in preparation of country ink.

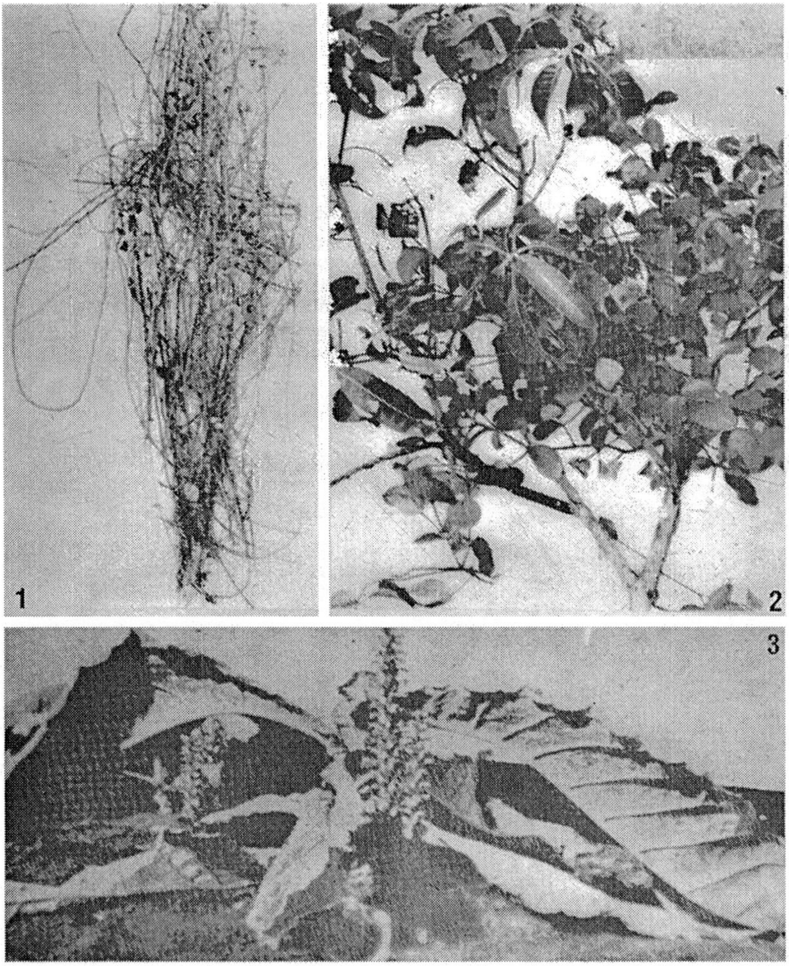

Plate 7: (1) *Cuscuta* spp. (Dodder), (2) *Loranthus* (On mango branch), (3) Orbanche

2. *Loranthus* spp.

Loranthus ....Loranthaceae

Loranthus is a phanerogamic parasite, found mostly in the tropics; rooting beneath the bark of the trees, particularly mango, it derives the whole of its nutrients from the juice of host plants and also receives certain stimuli. *Loranthus* hangs in clusters from the branches of trees, spreading roots on the bark.

Leaves are mostly opposite, simple and entire; stipules absent; scarlet flowers are bright, petals free, stamens are inserted at their base; ovary inferior ovules distinct; fruit is a berry which is carried away by birds and help its spread. Seed is solitary, devoid of testa.

Control measures

1. Affected branches of host plants should be periodically cleaned or cut down before monsoon.
2. High speed diesel will be helpful to kill the foliage of this parasitic weed.
3. Paraquat 1% spray will be helpful to kill the foliage of this parasitic weed.
4. Paraquat 3.5 1 plus 2,4-D at 3.5 1 mixed in 160 1 water/ha sprayed with a rocker sprayer on the infected area, can kill *Loranthus* by both contact and translocation action. This spray can be repeated but enough precaution has to be taken to protect the host tree.

3. *Orobanche aegyptiaca*

Broom rpe... Orobanchaceae
Hadda, (Hindi)

Orobanche spp. Have two common species. *Orobanche indica* and *Orobanche cernua. Orobanche cernua* is common in tobacco tracts in India which was recorded as early as 1883. it is an annual parasitic herb, a root parasite occurring in cultivated places on host crops like tobacco, brinjal and tomato, receiving certain stimuli and nutrition. It is a serious parasite on solanaceous plants which is favoured by high soil moisture. This parasite reduced shoot growth and even wilting of leaves of host plant. The host root-exudates, a chemical which stimulates the seed to germinate of Orobanche within 3 weeks which in turn forms haustoria on them. Soon after forming haustoria, this emerges through soils 5 to 6 weeks after planting, gets established as a total parasite and grow to a size of 20 to 45 cm high. It produces a swollen cloen connected to host roots by means of a fragile attachment through which it draws all its nourishment. Later, they develop several flowering shoots; flowers solitary in the axils of bracts, often crowded; each shoot produce numerous small seeds with fleshy endosperm in about 8 weeks. Seeds are disseminated by wind, birds, farm animals etc. It is propagated by seeds which can remain dormant in soil for several years.

Control measures

1. It is very difficult to control Orobanche because of its regrowth. Persistent removal as soon as it appears above ground and it seeds, is suggested. This method has to be continued 3 consecutive years.
2. Swabbing with paraffin gives quick knock down effect.
3. Glypohsate at 0.5 kg/ha is found effective.
4. Dalapon 6 kg or dlinuron 4 kg/ha cause selective control of Orobanche in tobacco and brinjal.
5. Spray of crag-herbicide 1 to soil, can reduce the incidence.

4. *Striga asiatica*

Figwort .... Scrophulariaceae
Agiya (Hindi)

Striga has two common species; *Striga asiatica* and *Striga hermonthica, S. asiatica* attacks sorghum and maize in India and *S. hermonthica* found in African countries.

Striga is a annual root parasitic crop-bound weed of enormous destructive power. It is well-known by its common name 'Witch-weed' because the parasitic nature of this weed is not well understood; while it is in bloom with attractive flowers.

Seeds never germinate in the season they are produced or when first shed; because they need a resting period. Seeds inside water in the following season but require a period of 'Sensitization' for 2 weeks and during this period, host becomes established. In response to a stimulant exuded by the host's root, the *Striga* seeds germinate and ensure its seedlings to make contact the stimuli for 24 to 48 hours. The exudates of sorghum roots contain mainly an active substance, lipophilic, similar to 'Strigol'.

Soon after, the seedlings attach itself to a compatible host within 96 hours and makes contact with a host root for its survival. Its tip swells and produces sticky hair like structure, known as pre-haustorium and gets established as a total parasite for first 3 weeks and when emerges as a green plant, it becomes a partial parasite. The *Striga* seedling anchors to its host and penetrate into the host roots vascular system to make xylem-xylem contact. Thus the haustorium acts as the parasite's umbilical cord; which is metabolically very active and depend entirely on their host for nutrients.

The parasite carefully maintains that part of the crop which provides it with mineral nutrients and water from the soil; but at the same time inhibits the development of the above ground parts of the crop, so as to reduce the host's need from the same sources.

The green active *striga* plants continue to rely on the host for carbohydrates, nitrate, sulphate and some essential amino-acids. It shows drought stress symptoms and high rate of transpiration, and thereby ensures the supply of food materials essential for its growth. Its stomata closes only partially at night and remain open when the water content of the leaf tissue falls below 20%.

Striga is an erect branching herb with linear rough leaves and yellowish tubular attractive flowers. Fruit is a cylindrical capsule with numerous seeds, 90,000 to a million per plant. Low soil fertility and low-rainfall favour its growth, causing more problems to subsistence farmers.

Control measures

1. In heavily infested areas, both parasitic weed and the crop should be ploughed down at early stages.
2. Rotating cereals with catch crops and hand-weeding can reduce the weed population considerably. Cotton, cowpea, groundnut, soybean etc. are effective trap crops.
3. Soil may be fumigated with methyl bromide keeping gas tight cover on the soil surface to rid the soil of *Striga* seeds.
4. Increased dose of nitrogen application may help good growth of host crop and suppress the parasitic weed to some extent.
5. Sorghum varieties resistant to *striga asiatica* can be evolved which produce no stimulant or which involves some form of mechanical obstruction.
6. Agronomic practices may be coupled with plant breeding programmes aimed at development resistant/tolerant lines.
7. Herbicidal control like spraying of 2,4-D (amine) at 1.2 kg/ha on *Striga* flowers or preplanting application of dicamba or fenac at 2 kg/ha are very effective.

CHAPTER 5

Aquatic Weeds

Aquatic weeds are those plants growing in or near water and complete at least a part of their life cycle in water resources. Any aquatic plant that causes adverse effects physical, chemical or biological on water-bodies, resulting in economic losses, is called aquatic weed. Aquatic weeds hinder navigation; chocking rivers, irrigation channels, lakes etc., impade drainage; interfere swimming recreation on water bodies; bathing and fishing; increase silt-deposition; harbour specific vectors of human diseases like mosquitoes. They cause enormous loss of water through transpiration from water surface. They also cause problems of irrigation by obstructing adequate supply at the distant fields from the water source by reducing the water flow by more than 50%. Decomposition of aquatic weeds depletes oxygen, thereby causing distress to fish culture.

In recent years, the rapid spread of aquatic weeds by vegetative and other means in tropical and sub-tropical regions, is creating serious socio-economic problems; causing enormous economic loss of water resources, some of the aquatic plants like water hyacinth which were imported for ornamental purposes earlier, have become serious problems throughout i.e. *Eichnornia crassipes* (water hyacinth), *Typha angustata* (Cattail), *Hydrilla verticillata* (Hydrilla), *Potamogeton* Spp. (Pondweed), *Salvinia molesta* (water fern), *Ipomoea aquatica* (Kalmi), *Nymphaea* Spp. (Water lily) etc.

Aquatic weeds (alphabetic order)

Scientific name	**Common name**	**Family**
Alisma plantago / triviate	Water plantain	Alismataceae
Alternanthera philoxeroides (Mart)	Alligator weed	Amaranthaceae
Alternantha logifolia, weed		Pontederiaceae
Anabaena spp.	Anabaena	Nostocaceae
Azolla pinnata, R. Br.	Azolla	Salviniaceae
Azolla imbricate, Waxai	Waterfern	-do-

Scientific name	Common name	Family
Chara zeylanica	Musk grass	Characeae
Coix aquatica	Jobs tears	Gramineae
Cladophora spp.	Cotton mat-type alga	Cladophoraceae
Commelina nudiflora	Dayflower, false flax	Commelinaceae
Commelina bengalensis L.	Kanchara, Kena	-do-
Cyperus corymbosus Rottb.	Small flower	Cyperaceae
Cyperus difformis L.	Umbrella sedge	-do-
Echinochloa colonum Link	Jungle rice, water grass	Gramineae
Eichhornia crassipes (Mart) Solms.	Water hyacinth	Pontederiaceae
Eleocharis plantaginea R. Br	Spike ruch	Cyperaceae
Fimbristylis miliacea Vahl	-	-do-
Heteranthera limosa (Sw) Wild.	Mud plantain	Pontederiaceae
Hydrilla vericillata Royle	Hydrilla	Hydrocharitaceae
Ipomoea hederaceae Jocky	Nil-kalmi	Convolvulaceae
Ipomoea aquatica Forsk	Kalmi	-do-
Jussiea repens L.	Water primrose	Onagraceae
Lemma minor	Duck weed	Lemnaceae
Marsilea minuta L.	Pepper work	Marsileaceae
Monochoria vaginalis Prese	Carpet weed	Pontederiaceae
Myriophyllum spicatum L	Eurasian water milfoil	Haloragaceae
Najas minor All.	Naiad, bushy pond weed	Najadaceae
Nelumbo nucifera C.	Lotus	Nymphacaceae
Nitella hyaline	Stone work	Characeae
Nuphar lutea	Yellow water lily	Nymphaeaceae
Nyphaca alba L.	White water lily	-do-
Paspalum fluitance	Water paspalum	Gramineae
Panicum purpurascens	Paragrass	Paniceae
Phalaris arundinaceae	Reed canary grass	-do-
Phragmites communis Trin	Common reed	-do-
Pistis stratiotes L.	Water lattuce	Araceae
Pithosphora spp.	Wet wool type algae	Chlorophyceae
Polygonum amphibium	Water smart weed	Polygonaceae
Pontederia cordata L.	Pickerel weed	Pontederiaceae
Potamogeton crispus L.	Curly leaf pond weed	Potamogetonaceae
Potamogeoton natans L.	Broad-leaf pond weed	-do-
Potamogeton pusillus L.	Small pond weed	-do-

Scientific name	Common name	Family
Sagittaria guaynesis H.B.K.	Arrow head	Alismataceae
S. pygmaca	Swamp potato	-do-
Salvinia auriculata Syn. – Molesta, Mitch	Water fern	Salviniaceae
Salvinia natans	Water fern	-do-
Scirpus articulates	Bulrush	Cyperaceae
Spirodela polyrhiza (L) Schleid	Giant duck weed	Lemnaceae
Spirogyra spp.	Slimy green alga	Chlorophyceae
Typha angustata	Cattail narrow leaf	Typhaceae
Typha latifolia L.	Common cattail	Typhaceae
Utricularia flexuosa Vahl	Bladder work	Lentibulariaceae
Vallisneria spiralis L.	Eelwwed, eelgrass	Hydrocharitaceae
Zannichellia palustris L.	Horned pond weed	Zosteraceae

The aquatic weed may be classified into broadly six groups:

1. Free-floating or surface species like water hyacinth
2. Rooted floating-leaf species like water lily
3. Shallow water submersed species like algae
4. Deep-water rooted submersed species like hydrilla
5. Emergent hydrophytes species like Typha
6. Amphibious marsh species like Kalmi

1. Free-Floating Species

These aquatic weeds float freely on the water-surface. They have majority of their leaves and flowers on the water surface. Some of them are attached to the sides of the ditch like Pistia (water lettuce), Lemna (duckweed) etc. the free-floating weeds cover large areas of stagnant water bodies, forming a large mat in association with each other. They also inter-mingle with submerged and emergent plants. They are capable of both vegetative and sexual reproduction.

2.Rooted Floating-Leaf Species

These aquatic weeds are firmly rooted on the bottom, but the aerial stem and leaves rise above the water surface. Their leaves may blanket shallow-water bodies, hinder navigation and fishing and prevent the use of water. These aquatic weeds include Potamogeton (Pondweed), Nymphaea (water lily) etc. The under-surface of these floating leaves, supports for egg deposition of snails, beetles, mosquito larvae etc.

3. Sub-Mersed Species

These weeds have most of their vegetative growth beneath the water surface and are rooted to the bed of ditches. They hinder fishing, block irrigation canals and reduce the flow of water. Some species like Hydrilla grow below the water surface. Fragments of brown submerged plants may float near water surface and continue its growth. They absorb the dissolved salts from water through their entire body.

Algae

Algae are shallow-water submerged weed which are primitive plants devoid of true leaves or seeds, reproduce vegetatively and also through spores. These are three types of algae: (i) Uni-cellular phyto planktons like Anabaena (blue-green alga); (ii) filamentous algae like cladophora (cotton mat type). Pithophora (wet wool type) and Spirogyra (slimy green type) and (iii) higher algae resembling vascular plants, like Chara (muskgrass) and Nitella (stone wort).

Methods of Aquatic Weed Control

Aquatic weeds can be controlled mechanically, chemically or biologically, the degree of control may vary with the situation.

A. Mechanical/Physical Control

Mechanical control of aquatic weeds means either physical removal manually or using simple implements / power-operated machines. This conventional method for eradication of aquatic weeds is free from any pollution causing; but this process is quite costly, laborious and time consuming. This can be practiced only in lightly infested, limited shallow water areas.

Use of mechanical weed-cutter becomes necessary in larger water-infested areas; chaining and dredging are two common mechanical methods of aquatic weed control. Later, under-water weed cutters and ditch-bank mowers, tractor-mounted or trailed sprayers etc. were introduced. However, different type of aquatic weeds viz. free-floating, submersed or emergent weeds create different problems in mechanical control.

a. Free-Floating Weeds

A small portion of weed-infested area encircled earlier by floating bamboos or ropes, is pulled ashore slowly by employing manual labour which is a slow and time-consuming process. For large areas, mechanical floating-weed harvester, saw boats etc. are used to cut the weed-mats into small blocks and to pull the weeds ashore through conveyer belts for manual removal.

b. Sub-Mersed Weeds

Sub-mersed weeds require repeated manual or mechanical operation at regular intervals to cut and pull them ashore; harvested weeds float to the surface and are collected manually or mechanically with sieve buckets. In a small and shallow-water area, long-tynedrakes with long handles are used for manual operation; but in larger water-infested areas, use of mechanical under-water weed cutters become necessary. Chaining is done for clearing ditches, tanks and river-beds, but this is very expensive and trained hands are required to handle it.

c. Emersed Weeds

It is not possible to eradicate them completely either by cutting with hand-tools or by use of power-operated machines, because of their rapid regrowth.

B. Chemical Control

Chemical control of aquatic weeds with effective herbicides is superior to mechanical control, as it does not require frequent operation and is comparatively easy, cheap and fast acting. Many effective aquatic herbicides have been screened to control both surface and emersed weeds; but for submersed weed control by introducing herbicides into the water, it has to be restricted because of possible pollution of water for its multi-facet uses. Aquatic herbicides should not be used indiscriminately as it may effect on aquatic ecosystem, create some potential risks to people using the water for various purposes. The degree of risk depends upon the toxic properties of the herbicides, the amount of chemical used and dangers of water contamination; time of treatment and type of application equipments used etc.

Estimation of Herbicides to be Used

The risk of aquatic herbicide use can be assessed, after the concentration of herbicide in the water is estimated, based on the active ingredient (ai) content in the chemical and the dilution factor involved. The dilution is given in ppm and is worked out by estimating the volume of water from the areas and depth as per formula

$$Wf = \frac{v \times c \times 100}{wa}$$

where

wf = weight of formulation required in gms

wa = weight of active ingredient in gm per 100 gm of formulation.

V = water volume in cubic meters

$$C = \frac{wf \times wa}{v \times 100}$$

C = concentration of active ingredient (ai) required in mg/L or g/cu.m.

When the concentration of ai of herbicide present in a body of water is determined, any dilution through inflows below the point of treatment, may be taken into account.

Selection of Herbicides

Before selecting a herbicide for use, it s effect on troublesome aquatic weeds; dose, method and time application; adverse side-effects to water users etc. have to be considered for its safe use, without any harmful effect. The risk of harmful side-effects is governed by

1. The property of herbicides ie. its toxicity, persistence etc. and
2. The condition and situation of the water course to be treated.

Chemicals used for Aquatic Weed Control

The aquatic weed control chemicals include copper sulphate acrolein, amitrole, dalapon, dichlobenil, diuron, diquat, fenac, paraquat, 2,4-D ec. For weed control in low-lyign rice fields, benthiocarl, C-288, machete, molinate etc. are used.

Copper sulphate is the oldest but most popular algicide, simazine is used in recent years for eth control of algae and also for soil sterilization on ditch banks at higher doses. Dichlobenial and fenac is used in exposed ditch bottom treatment; 2,4-D, paraquat and diquat to control floating and marsh species, diuron and endothal to control most species of Pond scumb; acrolein to control submersed species; amitrol and dalapon for ditch bank grasses, Cattails (Typha), bulrush sedges and reeds, silvex granule or silvex LV are used to control difficult surface and emersed weeds, occurring in confined waters.

Precautions

1. Only those herbicides recommended, to be used in water for aquatic weeds.
2. Water authority to be consulted before using a herbicide to avoid any pollution effect.
3. Method and rate of herbicide to be followed, as per instructions prescribed.
4. When applying to flowing water, operation to be proceeded up stream.
5. Precautions to be taken strictly while handling and applying aquatic herbicides.
6. Empty containers and surplus mixtures to be disposed safely.
7. Bathing, swimming, use of water by man or animal to be restricted for a specific period at least 10 to 15 days.
8. Wetting agents or additives to be avoided which are toxic to fish and wild life.

C. Biological Control

Biological control of aquatic weeds has attracted considerable attention in recent years. Fishes, snails, ducks and some insects are used in biological control; but has very limited use because of lack of precise knowledge of the controlling bio-agents. It is therefore necessary to screen and evaluate suitable biological control agents.

A suitable bio-agent can restore a desirable balance in the aquatic eco-system and can reduce the growth of aquatic weeds without polluting the water. Certain fresh water fish consume large quantities of aquatic vegetation. The common carp (*Cyprinus carpio*) and introduced grass carp (*Ctenopharyngodon idella*) are the natural control agent, particularly of submerged aquatic weeds like *Hydrilla verticillata*, *Potamogeton* spp., *Vallisneria spirallis* etc., while silver cap lives on planktons. The grass carp is found to be the most promising bio-agent for clearing submerged weeds.

Mantee (Sea-cow) and Hippos are also vigorous eater of water-plants and hence can be useful bio-agents in large canals and rivers. Buffaloes are also heavy eaters of aquatic plants.

Fresh water snail (*Marisa Cornuarietis*) feeds vigorously on several submerged weeds e.g. Pond weed, naiad, elodea and algae. It feeds on roots of water hyacinth, water lettuce and also on salvinia leaves.

Certain insects like flea beetle (*Agasicles hygrophila*) is a specific feeder of alligator weed. Its adults and larva feed on the weed leaves and then bore and pupate on left-over stems.

Aquatic Weeds

1. Algae

a. Anabaena spp. (Blue-Green Algae – Nostocaceae)

These are unicellular phyto-planktons which are microscopic, single cell, simple form, usually called water-bloom e.g. Blue-green algae.

b. Cladophora spp. (Filamentous Algae – Cladaphoraceae)

These are filamentous algae, irregularly branched. As dense growth occurs, gases are captured and hair-like strands colonies, to form scumbs and floating mats of alga. These scumb resemble as mass of cotton-mat type.

Other floating mats of filamentous algae are *Pithophora* spp. Resembles a mass of wet wool and Spirogyra spp. a slimy and green type alga. Spirogyra is a green filamentous unbranched alga floating in fresh water. These grow around shallow ponds and slow-moving streams.

Dense growth of all these filamentous algae prevents fishing depletes Oxygen, favours mosquito breeding site and gives an undesirable appearance.

c. (a) Chara zeylanica (Mass Grass – Characeae)
(b) Nitella hyaline (Stone Work – Characeae)

These are higher algae resembling vascular plants, anchored to the mud, possessing stems and branched symmetrically and encrusted with calcium carbonate deposits that gives a coarse feeling. These branched filamentous algae when crushed emit a musky odour similar to garlic and give a noxious small. Their dense growth may impede water-flow in canals and interfere with fishing.

Control measures

1. Copper sulphate is the most popular algaecide which can be applied 15 to 30 kg/cumec in every 5 to 10 km moving water. Water containing upto 2.5 ppmw, $CuSO_4$ is safe for human consumption and upto 10 ppmw for live stock drinking; however fish is sensitive to it beyond 1 ppmw concentration.
2. Simazine, diuron and monurno at 1 ppmw can be used as algaecides for ditch bank sterilization or for treating fish ponds. These are safe for recreation but unsafe for domestic

use for 2 to 3 weeks.

3. Acrolein at 50 ppm is the best to control higher algae like *Chara* and *Nitella*. Chlorxuron 6 kg, nitrofen 3 kg and SWEP 3 kg/ha are excellent herbicides to control algae.
4. Simazine at 1 ppm w is applied even to hatcheries for algae control.

2. *Alisma plantago*

Water plantain .... Alismataceae)

Alisma is a genus of hardy perennial aquatics which produce a stout proliferous corm, normally found in ponds and still water.

Leaves are usually cordate-ovate, thin and narrow; flowers small, white or pale.

Control measures

It is difficult to control because of the underground corm formation. However, 2,4-D mixed with paraquat may help to control this weed.

3. *Alternanthera phylloxeroides*

Alligator weed .... Amaranthaceae

Alligator weed is a vascular, very prolific aquatic noxious plant which grows mostly along the coastal region of the tropics, in areas of sewage inflow and fertile sites. It is a fleshy plant, forming dense mat of vegetative in shallow muddy places with stems that become prostrate.

Leaves are narrow, opposite, lance-shaped with a distinct midrib. Each node has two buds which sprout and produce shoots. Stems are hollow. It produces a short spike with white flowers, which are borne on long peduncles in a solitary head. It reproduces vegetatively through fleshy stem which is broken-off quickly, rooted easily and this spread rapidly on moist soil or submerged conditions. It interferes with flow of water in streams and canals.

Control measures

1. Alligator weed is very much resistant to herbicides and hence somewhat difficult to control. However, in shallow water, 2,4-D at 1 to 2 kg/ha may be applied and repeated application may be required for its control.
2. Paraquat or diquat at 6 kg/ha (2%) and silvex gr. 20 kg/ha (1% liquid) are very effective.

3. Biological control is an alternative to suppress this weed. *Agasicles hygrophyla* (flea beetle) larvae feed on leaves of this weed and bore into its stem to pupate inside and thus damage occurs. Other effective bio-agents are: *Anynothrips andersoni* (thrips) and *Vogtia molloi* (Stem borer).

4. *Azolla imbricate*

Water fern ....Salviniaceae

Azolla is a small genus of floating aquatic fern with small individual leaves, banded together in a scale-like manner. The upper portion of each leaf is floated while the lower portion is submersed. The species is a true fern and produces globular sporocarps. It is commonly found on shallow water, covers densely the water surface of the rice field and suppress the growth of rice plant. It covers entire pond or canal and interferes with fishing. Rootlets are unbranched and inconspicuous from the underside of the floating green leaves. It reproduces and multiplies vigorously by self-division in early summer at high temperature and grow readily in water containing little nutrient. It covers completely any quite water surface and thus useful in preventing the propagation of mosquitoes.

Control measures

1. PCP at 2.5 kg/ha has been proved very effective.
2. 2,4-D ester LV at 1% may be applied for its control.

Economic importance: *Azolla* is extensively used as bio-fertilizer by multiplying them and incorporating in the puddle soil before transplanting rice.

5. *Commelina bengallensis*

Kanchra, Water grass, Day flower .... Commelinaceae

Commelina is an annual, herbaceous, sub-succulent herb, found in damp places and rice field levies. It is an emergent hydrophyte with creeping stem, freely branched and rooting below from nodes. Root is fibrous, leaves long, sessile lanceolate and have a membranous sheath. Inflorescence consists of axillary cymes on peduncles. Flowers blue and small. Opening for a day; fruit is a capsule. It reproduces both by seeds and vegetatively and is mostly of easy culture.

Control measures

1. In rice fields, deep ploughing after kharif rice harvest and hand weeding in cultivated field in practiced.
2. In law-land rice, machete 5 l/ha (actual) as pre-emergence or stam F-34 5 l/ha (actual) as post-emergence is found to be the best.

6. *Cyperus difformis*

Umbrella plant .... Cyperaceae

The *Cyperus* spp. are annual sedge, most common in moist lands and low-lying rice fields. Those become predominant and their competition server in early stages under puddlled condition, cause the reduction in grain yield more than 50%.

The roots are fibrous and do not produce tubes; stems glabrous shiny, long without nodes; inflorescence is supported by leafy bracts with long spikes. These can easily be differentiated according to their inflorescence shape. They reproduce by seeds.

Control measures

1. Weed free condition upto 21 days after planting is sufficient for getting economic yield in transplanting rice. Hence pre-emergence application of machete granules 1.5 kg ai/ha (20 kg/ha) or Oxadiazon 0.75 kg/ha in combination of 2,4-D ester or C-288 (20 kg/ha granules) applied 4-6 days after transplanting is very effective.
2. Benthiocarb gr. 20 kg/ha (actual) or stam F-34 5 l/ha (actual) are also effective for control of these sedges.

7. *Eichhornia crassipes*

Water hyacinth .... Pontederiaceae
Jal kumbhi (Hindi)

Water hyacinth is a perennial, free-floating beautiful but notorious aquatic weed with hanging fibrous roots at the nodes, commonly found in stagnant water, ditches, ponds all the year round. It is often called a water orchid because of delicate colouring and structure. It is a native of tropical America, known as the famous "million-dollar weed", has now been naturalized almost throughout India. It is not considered as a most serious water weed and spreading with extreme rapidity. It is also partly fixed in mud, in shallow water and multiplies fast, forming a dense mat. It hampers navigation; impede flow of water in streams and channels, and interfere with fishing etc.

In the axils of leaves, a tuft of leaves and roots arise. The roots are fibrous, numerous, adventitious, branched which hang in water. When the connection between tuft and the parent plant gets broken, the off-set becomes a separate plant and multiplies fast. The leaves are 10-15 cm long, bright green, shiny spoon shaped, set across and upright. Petiole is often long, round, spongy and much swollen in the middle and these curious bladders facilitates easy free floating. The beautiful violet-blue flowers emerges from the centre of the rosette, showy arranging in dense spikes on long stalks. They grow rapidly and flower profusely throughout the season. It reproduces from off-sets, grows vegetatively @ 500 kg/ha per day and are inter-connected by stolons. It produces viable seeds as many as 5000/plant which remain viable in the mud about 10 to 15 years.

Control measures

1. It is difficult to eradicate this notorious aquatic weed. It is normally removed manually and confined by erecting fences around infested area. However, manual removal is laborious and costly.
2. Various mechanical devices have been designed for its collection and composting.
3. 2,4-D at 2 kg/ha (0.3%) can cause about 70% sink and the remaining by repeated application after 2 weeks, but this chemical control is to kill only the above aerial portion and thus-reinfestation completes within 6 months. However, the combination of 2,4-D + paraquat in equal doses at quite effective because of contact and translocative action simultaneously.
4. Paraquat 1 kg + amitrole 3 kg/ha or amitrole + fenac 0.1 kg ai/ha or dichlobenil 15 kg/ha are very effective. Amitrole alone at 1% spray is also effective.

8. *Eleocharis plantaginea*

Spike rush .... Cyperaceae

Spike rush is an annual sedge, found in low-lying areas, in shallow water ditches and rice fields. It can withstand a wide range of water conditions. It forms a dense mat from the small stems. The culms are 10-115 cm long, thread-shaped, often branched. Spikes found on culm terminal. It occurs in that profile of a lake where the photosynthetic biomass is concentrated just above the basal layer.

Control measures

Oxadiazon 0.75 kg/ha in combination with 2,4-D ester 1 kg/ha can control the rush very effectively.

9. *Elodea canadensis*

Flodea .... Hydrocharitaceae

Elodea is an aquatic ditch moss, hardy perennial plant, found in slow streams and ponds.

It is slender, wholly submerged plant, with branching upto 1 m long, according to depth of water. Flowering and fruiting take place below the water surface. The seeds raise and heavy buds drop to the bottom and grow in the spring.

Control measures

Biological control measures are preferable. Ducks and swans are found of it and hence controlling this through theses bioagents is the best method.

10. *Heteranthera limosa*

Mud plantain .... Pontederiaceae

Mud plantain is an annual herbaceous aquatic weed, found mostly in the muddy paddy field. It is an emergent plant with a fibrous root system; stem is round 20-40 cm tall; leaves ovate 8-10 cm long and smooth, having a long fleshy petiole. It produces violet flowers with 5 long petals; perianth with slender tube and black seeds. It is reproduced by seeds.

Control measures

Machete granules at 20 kg/ha (actual) applied in shallow water can control this weed effectively.

11. *Hydrilla verticillata*

Hydrilla .... Hydrocharitaceae

Hydrilla is a submersed aquatic weed rooted to the bottom forming large masses, leafy, dioecious. Since it slight requirement is extremely low, it grows in deep water upto 10 to 12 m. It is found in drainage channel, canals, ponds and lakes. Stems are long-branching which may break and form floating mats and grow very fast. The lower leaves are opposite and small; while the upper ones are in whorls and longer. Flowers arise from the spathe singly near the water surface on the growing tip; ovary extending beyond the spathe into a beak. It forms under-ground propagules and the reproduction is largely vegetative. It is most often used as egg collector of fresh water fish carps; which helps to collect eggs easily along with hydrilla and transfer to hatchery.

Control measures

1. Diuron (Karmex) 5 ppm, nitrofen (TOK-E-25) at 2.5 ppm, paraquat 5 ppm and 2,4-D ethy ester at 20 ppm –simazine 5 ppm is quite efficient for its control within 3 weeks period. At higher dose of 2,4-D and diuron at 1 to 2 kg/ha is found to be most phytotoxic and 100% mortality is expected. Paraquat 5 ppm is quickest in action and about 90% killing is assured.
2. Application of a mixture of diquat (1 ppm w) + cooper sulphate (4 ppm w) or fenac in combination with copper sulphate @ 0.05 to 1 ppm is found effective and economic treatment for successively controlling *Hydrilla* and *Vallisneria*.
3. As biological control measures, common snail (*Marisa cornuarietis*), grass carp (*Ctenopharyanodon idella*) feed upon *Hydrilla* and destroy them.

12. *Hydrocotyle umbellate*

Water pennywort .... Hydrocotylaceae / Umbelliferae

Water pennywork is a slender creeping perennial herb, rooting at nodes, thrive in moist places and abundant along irrigation channels. It is rooted in mud along pond and canal margins and grows into the water as a floating mat.

Numerous leaves are borne on the peduncle attached to the centre of round leaf. Leaves arise singly from each node on under water stems. Flowers very small, white appear in umbel-like clusters.

It interferes with flow of water in small drainage.

Control measures

It is difficult to control such water weeds.

Beneficial effect: Both leaves and stems are used as tonic and diuretic; as weak sedative; useful in diseases of nervous system and skin; decoction sued in leprosy treatment.

13. *Ipomoea aquatica*

Kalmi .... Convolvulaceae

Kalmi is an aquatic herb with hollow stem and provided with spongy floats. The plant is free-floating or creeping in moist places and irrigation channels.

Leaves are heart shaped, tri-lobed; flowers pink, pretty large and tubular.

Control measures

1. EPTC (Eptam) at 5 kg/ha can control it effectively. It is also susceptible to picloram (3 kg/ha) bentazon (1 kg/ha) prometryne (3 kg/ha) or MCPA (2 kg/ha).
2. Spray of 2,4-D 2 kg/ha may be useful.
3. Butachlor 2 kg, penoxaline 2 kg/ha can reduce the population of *Ipomea* considerably.

Economic importance: It is liked by buffaloes as succulent green feed and also consumed as 'Sag', a leafy vegetable.

14. *Juncos effuses*

Soft rush

Soft rush is an erect emergent rush that is neither a grass nor a sedge. It grows in swamps and prefers only little water where it can stand for prolonged period. It has a wide pH tolerance.

It grows in clumps of a few to several hundred stems, 1 to 1.2 m tall; clumps arise from a short stout rhizome which are palegreen, thick, soft and easily compressed. Inflorescence appears on the side of culm and the flowers borne on a peduncle.

It causes problem in small drainage ditches, impede waterflow and favour mosquito breeding.

Control measures

1. Oxadiazon 0.75 kg/ha 2,4-D ester 1 kg/ha may be effective to control soft rush.
2. Butachlor 2 kg/ha as pre-emergence is also effective.

15. *Jussiea repens*

Water primrose .... Onagraceae

Water primrose is a perennial herb, partly woody, found in wet areas, around aquatic gardens, ponds and in low-land rice, particularly in stagnant water. Once a flotant is formed, it begins to grow as a floating plant.

The roots are usually spongy and inflated. Stems erect, much-branched, glabrous; leaves floating, veined, ovate which grow on creeping stolons; flower yellow, showy and have long narrow calyx. Fruit is a cylindrical capsule, seeds are very small and brown.

Control measures

Diquat at 2 kg/ha and 2,4-D at 2 kg/ha can control this weed effectively.

16. (a) Lemma Minor (b) *Spirodela polyrhiza*

(a) Common duckweed ....Lemnaceae
(b) Giant duck weed ....Lemnaceae

Duckweed is a small, floating and simplest of flowering plants, found in stagnant ponds and sluggish streams. They are stem less and lack of true leaves; but have inconspicuous nerves, thick round fronds and number of rootlets. Small bulb lets are developed at the edge of the leaf which are detached and sink to the bottom and then germinate after the season. Very small flowers are born on a peduncle arising from the frond margin. They propagate from the base of the parent frond. Dense stands interfere with drinking of water by livestock; floating mass creates hindrance in small ponds and clog pumps.

The distinguished characteristics of giant duckweed are its 5 to 15 nerves and 4 to 16 routs; fronds are purplish on the underside. It is susceptible to wind movement.

Control measures

Chemical control with 2,4-D or 2,4,5-T @ 2 kg/ha can be done.

17. *Marsilea minuta*

Pepper wort .... Marsileaceae

Marsilea is an aquatic flowerless fern but looks like a flowering plant. It is normally fond in ditches, ponds, irrigation channels and rice fields. Stem consists of a creeping stolen with roots at the nodes. Leaves are erect, having long petioles with 4 leaflets.

Control measures

1. In direct-seeded rice, oxadiazon or butachlor at 1 kg/ha as pre-emergence can effectively control this weed.

2. Penoxalin 0.75 kg/ha on eighth day of planted rice is very effective.

18. *Monochoria vaginalis*

Carpet weed .... Pontederiaceae

Monochoria is an annual herbaceous semi-aquatic weed, mostly found in wet land rice fields in the tropics or sub-tropics. Root system is fibrous; stem fleshy, erect, 40-50 cm tall; leaves heart-shaped, long hollow petioled; oblong – lanceolate cordate, acute; inflorescence is a long spike; flowers blue; fruit is a capsule and seeds oblong. It is reproduced through seeds.

Control measures

1. Machete gr. 20 kg/ha or C-288 gr. 30 kg formulation/ha can control carpet weed very effectively.
2. Nitrofen(g) at 2 kg ai/ha, mollinate 2 kg/ha or oxadiazon 1.5 kg/ha are selective herbicides for its control.

19. *Myriohpyllum spicatum*

Water milfoil ....Haloragaceae

Myriophyllum is a troublesome perennial weed, most common in ponds and shallow water, aquaria pools. It is rooted in the bottom bed. It is grown submerged in water; but stem often emerges above water during summer; stem stont; lower leaves usually in whorls of 10 to 14, arranged in dissected segments; margin of emersed leaves are serrated. Each leaf resembles a small green parrot's feather with delicate foliage; composed of whorls of cut leaves; upper leaves resemble as bracts. Flowers are produced in the sessile of leaf axils. It is reproduced through both seeds and by vegetative propagation.

Thick growth affects shell-fish severely, prevents fishing and interfere boating.

Control measures

1. In small infested areas, it can be removed manually.
2. 2,4-D at 2 kg/ha is effective but repeated sprays at 3 weeks intervals is needed.
3. 2,4,5-T at 2 ppm gives effective control.

20. *Najas minor*

Naiad .... Najadaceae
Bushy pondweed (Hindi)

Najas has eight species, out of which Najas minor is most common in brackish ponds, lakes and canals. It grows as rooted submersed aquatic weed, can tolerate brackish water.

Stem is slender, stout, much branched and rooted at lower nodes, leaves slender, arranged in an opposite manner or crowded in whorls, margins have minute spines; bracts prominent; flowers dioecious and small seeds embraced in the leaf sheath. It is reproduced through seeds or by rooted stems.

Dense infestation in irrigation or drainage channels impede water flow more than 50% of the designed capacity.

Control measures

1. Acrolin at 50 ppm or dichlobenil 20 kg/ha is the best to control Najas, Chroxuron 5 kg/ha is very effective.
2. It is susceptible for 2,4-D diuron and sodium arsenite.

21. *Nelumbo nucifer*

Lotus ....Nymphaeaceae

Lotus is a large herbaceous aquatic plant, common in muddy shallow water tanks or large ditches. Stems are creeping on the ground under water and rooting at the nodes; have strong leaves, thick tuber bearing rhizomes, large, 40 to 60 cm long and round, long petioled either floating or mostly standing high above the water surface; leaf veins radiate from the centre; flowers big; white on red which are based on peduncles.

Control measures

2,4-D at 2 kg, diquat 2 kg or endothall 2 kg/ha is the best to control this plant.

22. *Nuphar lutea*

Yellow waterlily ....Nymphcaceae

Yellow water lily is an annual herbaceous aquatic plant, found in low lying areas, ditches, side of ponds and in paddy fields; grows well in sluggish canals, swamps, slow-running mud-bottom streams.

It is an emersed plant having creeping root stock; long petiole extends from root stock to floating leaf; leaves with prominent mid-vein; produces yellow flowers; petals mostly numerous, ripening of fruit takes place above the water and produces black seeds. It is reproduced through seeds. The plant spreads rapidly and curtails the water-flow seriously.

Control measures

1. Amino-triazole 8 kg/ha + 2,4-D (ester) at 6 kg/ha gives good control.
2. 2,4-5-T or Silvx at 1 ppm is found very effective.
3. Glyphosate 2 kg in 600 1 water/ha can effectively control water lily.

23. *Nymphaca alba*

White water lily ....Numphaeaceae

White water lily is rooted submersed perennial aquatic plant, mostly found in ponds and sluggish streams in shallow water.

Leaves are heart-shaped, emerged above water or float on the water surface; leaf-blades with prominent veins radiating from petiole attachment; flower large white, sweet- scented, showy, green on the outside, arises on a long peduncle.

Control measures

1. 2,4-D at 2 kg/ha or endothall at 2 kg/ha can control this weed.
2. Diquat at 1 ppm by surface spray cause raid kill of water lily.

24. *Panicum purpurascens*

Paragrass ....Paniceae

Paragrass is a perennial grass, grows most rapidly in moist soil, begins its growth along the banks and then spread out into the water. It is characterized by dense spreading mat growth. Stolons are 2 to 5 m long, collars densely pubescent with stiff hairs; panicles are long, distantly branched.

Dense stand can develop into floating mat which can create problems in drainage ditches to impede water flow.

Control measures

Repeated application of dalapon at 12 kg/ha after shoot emergence is effective.

Beneficial effect: It is very good green fodder for the cattle.

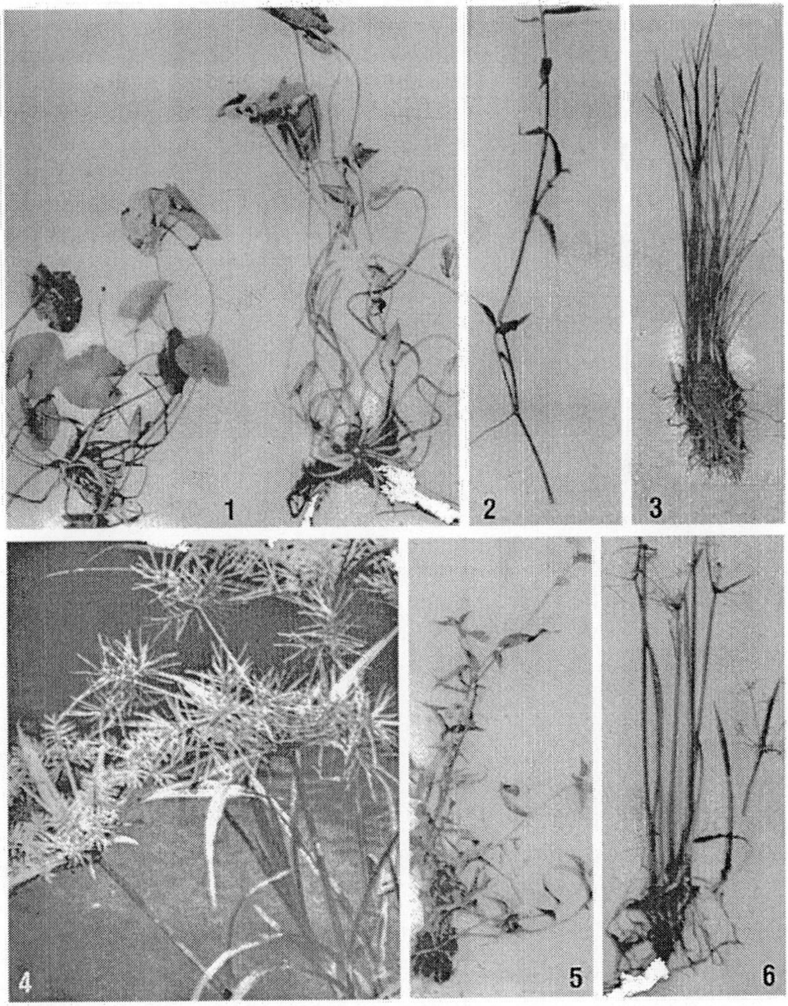

Plate 8: (1) *Nuphar letea* (Yellow waterlily), *Nyphaea stellata* (White waterlily), (2) *Commelina nudiflore* (False flax), (3) *Juncus effuses*, (4) *Cyperus diffuses* (Umbrella plant), (5) *Commelina bengalensis* (Water grass), (6) *Cyprus iria* (Rice sedge).

25. *Paspalum fluitans*

Water paspalum .... Gramineae

Paspalum is a perennial grass, mostly found along the banks in shallow water and in rice fields. It is a floating aquatic grass, forming large mats.

Stems elongate very fast, upto 3 cm per day; Thickening of epidermal cell-wall in the stem prevents penetration and translocation of the herbicides. Culms are glabrous or smooth which sprawl on water and form dense floating mats. They have stiff cores hairs on the loose sheath; inflorescence spreads just over water surface. It interferes with boat traffic and fishing on small streams.

Control measures

It is very difficult to eradicate by applying herbicide because of its peculiar body mechanism. Hence, repeated deep ploughing in the infested ditch bank areas is done for its control.

26. *Pharagmites communis*

Common reed .... Gramineae

Reeds are tall stout perennial grass with coarse, hard and long stems 3 to 4 m tall which resembles as a slim bamboo and long running root stock. It grows dense like a hedge along ditches marshes, border of the streams, wet places, swampy and shallow areas.

Leaves are flat, long with plume-like head, narrow long-tappering points. Ligule consists of a ring of hairs; panicles big, 30-40 cm, dark-purplish in colour; hairs extend from florets, give a silky appearance.

It spreads by means of stout creeping rhizomes which grow very deep and hence difficult to remove.

Control measures

1. Reeds are resistant to herbicides when growing in water. However, after draining of water, dalapon at 15 kg/ha + amitrol 2 kg/ha can be applied for its effective control.
2. Dalapon 19 kg + paraquat 0.5 kg/ha in 900 L water can control it effectively.
3. Glyphosate 2 kg in 600 1 water/ha can control this reed.
4. Aminotriazole 5 kg + dalapon 12 kg/ha is very affective.

27. *Pistia stratiotes*

Water lettuce, tropical duckweed............................. Araceae

Pistia is a free-floating aquatic tender weed found floating in stagnant shallow ponds and it has immense range throughout the tropics. Runners are called 'off-sets'; root-stalks touches the water column and produces long slender feathery numerous hairy roots. Light yellow green leaves in the form of a rosette is the main characteristics of water lettuce; leaves are wedge-shaped, spongy, inflated triangular in shape, 5-10 cm long, having many elevated ridges running lengthwise underneath; produce a large number of small, inconspicuous flowers in the centre of the plant. Vegetative reproduction is mainly true buds; stlons are easily broken and multiplied.

The vegetation interferes with drainage, render the water area useless; affected pond is often attacked by insects and becomes normally the breeding site of mosquitoes.

Control measures

1. Acrolein at 50 ppm is very effective to control water lettuce.
2. Diquat at 2 kg/ha and 2,4-D at 2 kg/ha can give a good control of Pistia.

28. *Pontederia cordata*

Pickerel weed .... Pontederaceae

Pickerel is a perennial aquatic weed, found mostly in shallow ditches, logs, ponds and low-lying rice areas. It is grown in clumps. It has creeping strong horizontal root stalk that has cluster of erect long petioled leaves, 8-12 cm weed, and terminal spike of violet blue flowers. It is propagated mostly by easy division.

The vegetation interferes with drainage and render the water area useless.

Control measures

Butachlor gr. at 20 kg/ha can control this weed effectively.

29. *Potamogeton* spp.

Pondweed .... Naiadaceae/ Potamogetonaceae

Pond weed is a noxious aquatic plant which has a stout under-water rhizome in muds attaching themselves to the bottom; mainly found in canals, streams, lakes and also getting serious in rice fields. Leaves are both submerged and floating; sub-merged leaves usually crinkled along the margin; spike borne at the stem tip, dense and cylindrical, 3-8 cm long. Its infestation hinders stream flow and boat traffic.

Control measures

1. Paraquat at 5 ppm concentration results its complete control.
2. Acrolein at 50 ppm, dichobenil 15 kg, chloroxuron 5 kg, diquat 2 kg or fenac 1 kg/ha can control this weed effectively.

30. *Sagittaria guayanensis*

Arrowhead .... Alismataceae

Arrowhead is a perennial hardy aquatic weed, found in swamps, shallow water, muddy places and drainage ditches. Stems have milky juice; leaves erect, lance-shaped, 5-10 cm wide, natural look like a arrow head. Leaves grow from the rhizome and tubers are produced at the end of rhizomes. White butter cup like flowers arise from the peduncle. It forms undesirable vegetation in shallow ditches.

Control measures

1. Since the leaves are oily, a good wetting agent has to be added in the chemical like 2,4-D or 2,4,5-T as foliar spray.
2. Endothall at 2 kg/ha is very effective.

31. *Salvinia* spp.

Water fern .... Marsiliaceae / Salviniaceae

Salvinia molesta is a free-floating aquatic weed, grows profusely in shallow ponds, canals, still-water areas and low-lying rice fields. It favours to grow best in organic areas in association with alelgator weed and water hyacinth. It is signified as noxious weed in the tropics, spreading fast and causing severe problems, particularly in low lying rice cultivation, water transport, irrigation, public health and thus the extend of damage is very high.

It looks much like a flowering plant but it is a spore plant. These spores are produced in small oval bodies known as sporocarps. After the plants die, the spore-capsules remain in the soil.

Its leaves are small, bright green, arranged in common stem, covered with bristle-stiff hairs and appearing folded in the centre, mid-rib is distinct. Sporocarps are formed on portions of the submerged leaves. It is propagated by vegetative means.

Other species like *Salvinia auriculta* has two forms, a mat form and on open-water form; while *Salvinia natans* is found in Kashmir area which forms extensive mat on the water-surface, changes physico-chemical characters of water, hinders navigation and recreation activities.

Control measures

1. Granular herbicide, C-288 at 30 kg formulation/ha (0.75 kg ai/ha) can completely control *Salvinia* spp. at low water level in rice field or low-lying areas, without any harm to standing rice crop.
2. For the open area, 2,4-D at 3.5 l/plus paraquat at 3.5 litre can be sprayed; however this process has to be repeated 2 to 3 times at 15 days interval for its complete control during the season.
3. Diquat at 2 kg/ha may be sprayed on the floating surface of *Salvinia*.

32. *Scirpus articulatus*

Bulrush .... Cyperaceae

Bulrush is an aquatic sedge, found in shallow ponds, stream banks which grow in the mud. It has the characteristics of straight, non-branching and cylindrical soft stem. Stems are long 2.5 to 3 m tall. Gradually tapering to the pointed tip, spongy and hence can be compressed easily. Panicle is formed at tip of the stem; flowers perfect, solitary, spikelets scaly and brown. Large colonics impedge the flow of water in shallow-drainage.

Control measures

2,4-D at 2 kg/ha and dalapon 5 kg/ha can effectively control this weed.

33. *Typha* spp.

Cat-tail .... Typhaceae

Two species of cattails are of most common.

Occurrence: *Typha angustifolia* (narrow leaf) and *Typha latifolia* (Common cattail)

Typha is an erect hardy perennial swamp-growing grass-like emersed weed with joint less stem which is common, mostly in marshy lands forming colonies and also in ponds, ditches, drains in both fresh and brackish water.

It is tall, grass-like with fleshy leaves (5 cm wide and 5 cm long) having no mid-vein. It is very gregarious and robust grass, 2 to 4 m tall; multiplied rapidly by underground rhizomes and through air-borne seed propagation. It produces more than 100 aerial shoots in one season and seeds remain viable for 5 years.

The rhizomes can go upto 60 cm depth and establish firmly. It has very dense. Cylindrical spike 20 to 50 cm long, resembles the tail of a cat and so the name. It contains pistillate in the lower portion and the staminate in the upper portion. Mosquito breeding is high in stands of cattail.

Control measures

1. Manual or mechanical weeding is very difficult because of its deep establishment and quick regeneration from underground rhizomes. However, repeated mowing at monthly interval and submergence for about 4 weeks can suppress this weed.
2. In shallow-water areas, dalapon at 15 kg ai/ha + amitrole at 3 kg/ha (0.2%) with 0.5% surfactant can give a good control.
3. Diuron at 15 kg/ha + MSMA 2 kg/ha can give conspicuous control.
4. A combined spray of 2,4-D amine + aminotriazole at 1.5 kg each/ha or amitrole 1% spray is very effective.

34. *Utricularia vulgaris*

Bladderwort .... Lentibulariaceae

Bladder wort is an aquatic plant, commonly found floating in fresh water and in mixed stands with other submerged plant. Its stem float just below the water surface and the free-floating nature creates problems to control as it moves from infested area to weed free area by water or wind movement in a short time.

Stems are branched and leafy; bladders numerous and conspicuous which are attached to the fine leaves; bladder trap small aquatic insects which enter through valve-like door for food but cannot escape. True leaves absent; flowers yellow, borne on erect scapes, 10 to 30 cm tall. It impedes water flow in a canal as it grows in mixed stands; harbours mosquitoes in the foliage, creates problems in their control because of their quick movement in water or wind.

Control measures

Diquat at 2 kg/ha and acrolein at 50 ppm are the best to control bladder wort.

35. *Vallisneria spiralis*

Eel-grass or tape gras .... Hydrocharitaceae

Eelweed is a submerged grass-like perennial plant with ribbonlike leaves, 1 to 2 m long resembles a grass in general appearance. It grows well in quiet fresh water, mostly in lakes or slow-moving streams, not more than 3 m deep.

Leaves are attached in clusters to a basal node of the rhizomes, upto 3 m long; and upper part of the leaf floating on the water surface. Leaves have netted veins; roots fibrous. Dioecious flowers are borne at the tip of a long peduncle with coils and retracts the fruit below the water after fertilization. Fertilization mechanisms is peculiar male flowers become detached from the plant and float to water surface and the female flower project above water surface and then fertilization takes place. Soon after the female flower goes down below the water surface and develop into berrylike fruits containing numerous seeds. It interferes with boating and fishing.

Control measures

1. Endothall at 1.5 kg/ha or dichlobenil at 5 kg/ha can control eelweed effectively.
2. 2,4-D and sodium arsenite are also effective.

Major aquatic weeds under different family

	Family	Scientific name	Common name
IA	Free-floating species		
	Araceae	*Pistia stratiotes* L.	Water lettuce
	Lemnaceae	*Lemna minor*	Duck weed
		Spirodela polyrhiza (L) Schleid	Giant duck weed
	Pontederiaceae	*Eichhornia crassipes* (Mart) Solnes	Water hyacinth
	Salviniaceae	*Salvinia auriculta* Mitch Syn. molests	
		Salvinia natans	-do-
		Azolla imbricate Waxai	Water velvet
B	Floating macrophytic Polygonaceae	*Polygonum amphibium* L.	Water smart weed
II Rooted floating leaf form species			
	Alismataceae	*Sagittaria guavanesis* H.B.K.	Arrow head
	Convolvulaceae	*Impomoea hederaceae*	Nil-Kalmi
	Nymphacaceae	*Nelumbo nucifer* G.	Lotus
		Nymphaea alba L.	White water lily
		Nuphar lutea L.	Yellow waterlily
	Potamogatonaceae	*Potamogeton crispus* L.	Curly-leaf pond weed

	Family	Scientific name	Common name
		Potamogeton natans L.	Broad-leaf pond weed
		Potamogeton pusillus L.	Small pondweed
	Zosteraceae	*Zannichellia palustris* L.	Horned pondweed
III. Shallow-water submersed species			
	Algae		
	Phytoplankton		
	Nostocaceae	*Anabaena* spp.	Blue green alga
	Filamentous algae		
	Cladophoraceae	*Cladophora* spp.	Cotton mat type alga
	Chlorophyceae	*Pithospora* spp.	Wet wool type alga
		Spirogyra spp.	Slimy green alga
	Higher algae		
	Characeae	*Chara zeylenica*	Musk grass
		Nitella hyaline	Stone wort
	Higher plant		
	Najadaceae	*Najas minor* all.	Naiad
		Valisneria spiralis	Eelweed
IV. Deep-wate rrooted submersed species			
	Halorhagaceae	*Myriophyllum spicatum* L.	Eurasian water milfoil
	Hydrocharitaceae	*Hydrilla verticillata* Royle	Hydrilla
		Eladea canadeusis	Eladea
	Lentibulariaceae	*Utricularia flexuosa* Vahl.	Bladder wort
V. Emergent hydrophytes species			
	Commelinaceae	*Commelina bangalensis* L.	Kanchra, water-grass
	Cyperaceae	*Cyperus difformis* L.	Umbrealla plant
	Gramineae	*Phragmites communis* Trin.	Commonreed
	Alismataceae	*Alisma plantago*	Water cattail
	Pontediriaceae	*Pontederia cordata* L.	Pickerel weed
	Typhaceae	*Typha angustata*	Cattail narrow leaf
		Typha latifolia	Common cattail
VI. Amphibious marsh species			
	In temporary water		
	Amaranthaceae	*Alternanthera philoxeroides* (Mart) Griseb.	Alligator weed
	Gramineae	*Panicum purpurascens* Raddl.	Para grass
		Paspalum fluitans Kunth.	Water paspalum
	Marsileaceae	*Marsilea minuta* L.	Pepper wort

	Family	Scientific name	Common name
	Pontederaceae	*Heteranthera limosa* (SW) wild	Mud plantain
		Monochoria varinalis Presl.	Carpet weed
	On clay sub-stratum		
	Cyperaceae	*Fimbristylis miliaceae* Vahl	Hoorah grass
	Forming floating mat		
	Convolvulaceae	*Ipomoea aquatica* Forsk	Floating morning glory, kalmi
	Hydrocotylaceae	*Hydrocotyle umbellate* L.	Water pennywort
	Onagraceae	*Jussiea repens* L.	Water primrose
		Ludurga parviflora	Water purselane
	Trapaceae	*Trapa bisponosa* Roxb.	Singhara

Herbicides for Aquatic Weeds

Weed species	Effective herbicides (kg or l/ha)
Algae	Copper sulphate (2.5 ppm w), Simazine Diuron or Monuron (1 ppm w), Acrolein (50 ppm w), Chloroxuron (6), Nitrofen (3), SWEP (3),
Alisma plantago	2,4-D (2) + Paraquat (2)
Alternanthera phylloxeroides	2,4-D (2), Paraquat (6), Diquat (6), Silves 1%, 20 g.
Azolla imbricate	PCP (2.5), 2,4-D ester 1%
Cyperus spp.	Machete g (20), C-288 g (20), Benthiocarb g (20)
Eichhornia crassipes	2,4-D (2) + Paraquat (2), Amitrole (3) + Fenac (1)
Eleocharis plantaginea	Oxadiazon (0.75) + 2,4-D (1)
Heteranthera limosa	Machete g (20)
Hydrilla verticillata	Diuron (5 ppm w), Nitrofen (2.5 ppm w) Paraquat (5 ppm) + 2,4-D (20 ppm) + Simazine (5 ppm), Diquat (1 ppm) + $CuSO_4$ (4 ppm), Fenac (1) + $CuSO_4$ (ppm)
Ipomoea reptans	Eptam (5), Picloram (3), Bentazon (1), MCPB (2), 2,4-D (2)
Juncos effuses	Oxadiazon (0.75), 2,4-D (1), butachlor 92).
Jussiaea michanxiana	Diquat (2), 2,4-D (2)
Lemna minor	2,4-D (2)
Marsilea minuta	Oxadiazon (1), Penoxalin (0.75)
Monochoria vaginalis	Machete g (20), C-288 g (30), Nitrofen (2), Molinate (2)

Weed species	Effective herbicides (kg or l/ha)
Myriophyllum spicatum	2,4-D (2), 2,4,5-T (2 ppm)
Najas minor	Acrolein (50 ppm), Dichlobenil (20), Chloroxuron (5)
Nelumbo nucifer	2,4-D (2), Diquat 92), Endothall (2)
Nuphar lutea	Amino-triazole (8) + 2,4-D (6), Silvex (1ppm), Glyphosate (2)
Nymphaca alba	2,4-D (2), Endothall (2), Diquat (1 ppm)
Panicum purpurascens	Dalapon (12)
Phragmites communis	Dalapon (15) + Amitrol (2), Dalapon (15)+ paraquat (0.5); Glypohsate (2), Aminotriazole (5) + Dalapon (12)
Pistia strtiotes	Acrolein (50 ppm), Diquat (2), 2,4-D (2)
Pontederia cordata	Butachlor g (20)
Potamogeton spp.	Paraquat (5 ppm), Acrolein (50 ppm), Dichlobenil (15), chloroxuron (5), Diquat (2), Fenac (1)
Sagittaria guayensis	Endothall (2)
Salvinia molesta	C-288, g(30), 2,4-D (3.5) + Paraquat (3.5), Diquat (2)
Scirpus articulatus	2,4-D (2), Dalapon (5)
Typha spp.	Dalapon (15) + Amitrole (3), Diuron (15) + MSMA (2), 2,4-D (1.5) + Amino triazole (1.5)
Utricularia flexuosa	Diquat (2), Acrolein (50 ppm)
Vallisneria spiralis	Endothall (1.5), Dichlobenil (5)

CHAPTER 6

Allelopathy in Relation to Weeds

Introduction

The term 'allelopathy' refers to biochemical interactions, both inhibitory and stimulatory, between different types of plants. Organic compounds that are inhibiting at some concentrations, are stimulatory to the same processes in very small concentrations. However, 'allelopathy' is mostly defined as any direct or indirect harmful effect by one plant or another through production of chemical compounds that escape into the environment, which leads to depletion of some essential factors like water, minerals food and light.

The allelopathic effects of weeds on crop plants in overall influence of one plant on another, may be termed as 'weed-crop interference', eliminating the possible term "weed-crop competition". The interference of several important weeds on crop involves reduction in yield ranged from 40 to 100%. Soil sickness problem might be due to exudates of weeds that injure the cultivated crop plants. Allelopathic potential of individual species of weeds (both, monocots and dicots) on crop yields has been demonstrated.

A. Interference of Monocot Weeds on Crops

1. *Agropyron repens* (Quack Grass)

Agropyron repens (Quack grass) is an important weed of field crops which causes serious decrease in yield of maize and potato. It interferes with uptake of minerals particularly nitrogen and potash by maize; and reduces the production of fresh and dry matter and also phosphate uptake of wheat even adequate amount of available P present in the soil.

Danks *et al.* (1975a) found that certain allelopathic agents significantly decreased uptake of minerals. Quack grass produces toxin which interfere with growth of crop plants. Ethylene is generated in the presence of quack grass rhizomes due to microbial activity in soil, which is partially responsible for 'alleged' allelopathic effects of the weed, resulting in decreased uptake of the minerals by associated crop species.

2. *Avena fatua* (Wild Oat)

Wild oat is a serious weed of winter annuals like what, barley and oats. Wild oat residues inhibit germination of certain herbaceous species in shrub stands due to an allelopathic mechanism. Growth of leaves and roots of wheat is significantly reduced by root exudates of *Avena fatua*. Two allelopathic compounds, typical of scopoletin and vanillic acid, have been isolated from its roots exudates.

3. *Cynodon dactylon* (Bermuda Grass)

Barmuda grass is a troublesome perennial creeping weed, found in cultivated lands. Decayed Bermuda grass residues remained in the field, inhibit seed germination, root and top growth of barley due to allelopathic effect. Growth inhibition is proportional to the concentration of plant material is the soil.

4. *Cyperus esculentus* (Yellow Nut Sedge)

Yellow nutsedge is a perennial sedge, infesting grain crops, soybeans, orchards, lawns etc. It inhibits the germination of oat and other crop species; inhibitory to root and shoot growth of maize and soybeans. The effect on soybean is considerable due to the allelopathic compounds, produced by yellow nut sedge. James *et al.* (1973) identified –hydroxybenzoic, vannillic, syringic, ferulic and p-coumaric acid in the yellow-nut sedge extracts.

5. *Cyperus rotundus* (Purple Nutsedge)

Purple nutsedge is a most persistent weed of cultivated fields. It reduces crop yields ranging from 23 to 89%; reduction in yield of rice about 38% in its high densities and water extracts of tubers inhibits seedling growth in all crop species with a maximum of about 85%.

Soil previously infested with purple nutsedge for 6 to 12 weeks, can reduce germination of mustard, barley and cotton seeds; and marked reduce the survival of radish, onion and tomato seedlings. The inhibition is greater in sandy soil than in clay soil. This sedge competes extensively with rive for moisture and transmission of light at its early stage.

6. *Digitaria sanguinalis* (Crab Grass)

Crab grass is an annual herbaceous grass, common in cultivated fields and roadsides. The interference of crab grass on yield reduction of cotton and groundnut ranges from 42 to 95%, and further Robinson (1976) reported no yield of seed cotton in 2 out of 3 years when crabgrass was allowed to remain in the rows. Water extracts of various organs of carb grass is quite toxic to several associated species, inhibit seed germination, inhibitory to growth of certain nitrogen fixing and nitrifying bacteria. Root exudates of crabgrass reduce the growth of seedlings of cultivated species and decaying roots inhibit the growth of maize roots.

7. *Echinochloa crusgalli* (Barnyard Grass)

Barnyard grass is the most troublesome weed in low-land rice during kharif season. It interferes with crop yields, the yield reduction ranged from 19 to 79%, depending on the crop and its environmental conditions. Water extract of barnyard grass seeds can significantly reduce seed germination of crop plants, inhibit hypocotyls growth of soybeans by 2.5 to 49% and height of maize by 5 to 24%. The allelopathic potential of this weed and its interference on crop yields are quite significant.

8. *Eleusine indica* (Goose Grass)

Goose grass is heliophytic and hence suppressed by reduction in sunlight intensity. Decomposing residue of this grass inhibits seed germination and growth of many species, particularly in sandy soil.

9. *Lolium multiflorum* (Italian Rye Grass)

Rye grass is commonly grown in wheat field of sub-tropical region. It suppresses germination and growth of many species in its vicinity. Decomposing residues of this weed is toxic to seedlings of oats, lettuce and sweet clover.

10. *Setaria glauca* (Yellow Foxtail) and *S. faberii* (Giant Foxtail)

Foxtail is an annual noxious weed, found in cultivated fields of maize etc., plantation crops and roadsides. Interference by this weed have an appreciable reduction on yield of maize, sorghum and soybeans; reduction being 28% in soybean and averaged 15%

when soil moisture is adequate. This weed has allelopathic potential. Decaying residues in soil of yellow foxtail significantly reduce the root growth and also the height of maize by 50% and growth of soybean by 25% to 49%. Aqueous extracts of yellow foxtail seeds reduce germination and inhibit radicle and coleoptile growth of maize and soybeans.

11. *Sorghum halepense* (Johnson Grass)

Johnson grass is a persistent perennial worse weed in sugarcane, maize, soybean etc. It cause an average yield reduction of 34 to 87% and its interference reduced soybean yield by 24 to 42%.

Root exudates and decaying residues of Johnson grass can inhibit both root and shoot growth or barley significantly.

B. Interference of Dicot Weeds on Crops

1. *Amaranthus dubius* (Pigweed)

Amaranthus is an annual weed in the tropics. This weed inhibits germination of seeds of crop plants like maize, sorghum and beans. Inhibition of soybean growth is found to be high, growth inhibition is more pronounced in sandy soil.

2. *Ambrosia artemistifolia* (Commno Ragweed)

Root exudate and also aqueous extract of shoot of the ragweed inhibit seed germination and seedling growth of radish, sorghum and other crops. The morphological change may be mediate by allelopathy. Decomposing leaves and roots are highly inhibitory to most crop species. Five phytotoxins have been isolated from this weed which reduce weed germination of radish and sorghum.

3. *Boerhaavia diffusa* (Spreading Spiderling)

Spiderling is a spreading perennial herbaceous weed, found in sugarcane etc. and wastelands. The extract of this weed affects the growth of seedlings of bajra, til etc. thus showing its allelopathic potential.

4. *Celosia argentea* (Cock's Comb)

Celosia is an annual plant, found in cultivated fields which is inhibitory to root and shoot growth of bajra seedlings. The exudates of young weeds (upto 25 days) are most inhibitory; inhibition decreases with age of Celosia seedlings. Root and leaf extracts of this weed is more inhibitory than stem extracts.

5. *Chenopodium album* (Lambs Quarters)

Lambs quarters is an annual weed of rabi season crops like wheat, pea etc. It reduces sugarbeet yields as much as 94% and the sucrose content by 63%.

Aqueous extracts of this weed can markedly reduce germination, Coleoptile root length of wheat, maize and soybean. Root exudated retard root growth of maize-seedling. Toxin from this weed reduces mitotic activity in roots of wheat and rye.

6. *Cirsium arvense* (Canada Thistle)

Canada thistle is a prickly winter perennial weed. It cause 40% reduction in yield of wheat. Water extract of its roots reduce 87% seed germination and radicle growth of clover in 72 hours. While extracts of foliage can reduce germination by 14%. Relative effects of allelopathy and interference of Canada thistle on crop yield have been determined. Toxins of *Cirsium arvense* also may reduce mitotic activity in roots of wheat and rye.

7. *Cucumis trigonus* (Wild Cucumber)

Cucumis is often climbing by tendrils found in fields of maize, sugarcane etc. water extracts of this species affect growth of seedlings of bajra, til etc. as it is the most stringent inhibitor.

8. *Convolvulus arvensis* (Field Bindweed)

Field bindweed is a twining deep-rooted perennial herb. Soils from patches of bindweed inhibit seed germination and seedling growth of species like maize, sugarcane etc. associated with it.

Allelopathic effects of bindweed on associated species cause detrimental effect in crop plants and its dominance for several years follows abandonment.

9. *Digera arvensis* (Digera)

Digera is an annual kharif season weed. It produces toxins that inhibit seed germination and reduce dry weight increment of bajra.

10. *Euphorbia* spp. (Spurge)

Euphorbia spp. is an annual weed, grown in cultivated fields of wheat, pigeonpea etc.

Water extract of this species inhibit germination and seedling growth of wheat. It exerts allelopathic activity against some weeds. Several phenolic compounds have been identified that are produced by the allelopathic species, Euphorbia supine.

11. *Melilotus alba* (White Sweet Clover)

Sweet clover is an annual rabi season weed. Soaking maize seeds in an aqueous extract of sweet clover for 24 hours can reduce seed germination and growth of maize roots.

12. *Partheium hysterophorus* (Ragweed, Congress Grass)

Ragweed is a pernicious troublesome weed which causes health hazards. Its interference with crop growth is due to allelopathic effect.

Aqueous extract of leaves, stems and roots decrease seed germination, seedling growth and chlorophyll production of radish, cabbage, cauliflower, groundnut, blackgram etc. Allelopathic potential of this weed has been clearly observed. Growth of plants associated with this weed is retarded; nodules of legumes reduced, particularly beans. Decaying leaves inhibit nodulation, growth and yield of beans, cowpea, tomato, ragi etc. Inhibits released o the soil from leaves and stems remain active for about a month. However, maximum exudation of inhibitors from its roots occur at its flowering stage.

13. *Phalaris* spp. (Canary Grass)

Aqueous extracts of its roots and tops cause twisted roots, prevents root hair development.

14. *Polygonum* spp. (Knotweed)

Knot weed is pernicious weed in lawns which affects Cynodon dactylon very badly in lawns. It reduce germination of cotton. Root exudates retard root growth of sorghum, top growth of bermudagrass and lambsquarters.

15. *Portulaca oleracea* (Purselane)

Purselane is an annual succulent herb. Water extracts inhibit seed germination and root growth of wheat.

16. *Rhus copalina*

Gas is produced during decay of its leaves which inhibit germination of radish seeds. The inhibition is due to the presence of volatile organic compounds as well as methane and carbon dioxide.

17. *Salsola kali* (Russian Thistle)

Six phenolic compounds have been identified in leaf extracts of this weed, which reduce germination and root growth of radish.

18. *Sonchus arvensis* (Sowthistle)

Sowthistle is a winter season weed, found mostly in wheat fields. Toxins of sowthistle reduce mitotic activity in roots of wheat, rye etc.

CHAPTER 7

Weed Control in Crops

Both the crop and weed belong to the same plant kingdom, sometimes the same genera and species, hence for the successful use of herbicides, the physiology of crops and weeds should be known clearly. Rainfed crops suffer badly due to rapid growth of many weeds; cotton, groundnut, sugarcane, etc. suffer most due to very slow germination and vegetative growth; weed control is a major problem in growing legumes and nursery stocks; heavy losses occur due to weeds in vegetable crops etc.

Though hand weeding and inter-cultural operations mechanically are the efficient method in certain row crops, mechanical weeding at early two weeks, followed by herbicides, application at 3 to 4 weeks stage can keep the crop weed-green. Certain general principles for herbicide application have to be followed:

1. Rapid growing, tender weeds can be killed most easily by herbicides than old ones.
2. Uniform distribution of chemical is essential in soil application and mechanical incorporation is desirable for its easy movement in the root zone.
3. Light sandy soil requires less chemical than clay or organic soils.
4. Over-doses have to be avoided and spray pressure should be kept minimum with translocated hormone-like chemicals like 2,4-D.
5. Soluble salt formulations should not be applied during rainy period.
6. Spray chemicals become toxic at high temperature, hence high temperature and high wind velocity should be avoided.

Selection of Crop Varieties for Tolerance of Herbicides

The genetic variation for tolerance of certain herbicides has been observed in many crops which will lead to the selection of specific strains and resistant to specific herbicides. Since selective herbicides have already been identified to suit major crop plants, emphasis is now to be given to select suitable cultivars for which potential of existing herbicides along with their known toxicology is needed.

Both genetic engineering and tissue culture techniques may be employed to stimulate the development of herbicide resistant cultivars which may be cheaper than to develop new herbicides for existing cultivars. Cultures of crop plants can be tested in a medium containing herbicides and the survivors could be propagated and tested agronomically. It may thus be possible to produce herbicide tolerant plants by transferring the resistant genes.

Selectivity of 'terbutryn' to wheat cultivars, Sonalika and Arjuna; atrazine to hybrid sorghum CSH-1; and benthiocarb to rice cultivars, IR-20 and GEB-24 are the specific instances.

A. Weed Control in Cereals

1. Rice (*Oryza sativa*) Gramineae

Reduction in yield of rice due to weeds, depends on the weed species and density of weeds. Among the weeds, barnyard grass (*Echinochloa crusgalli*) causes the greatest damage to dryland rice; while aquatic weeds like carpet weed (*Monochoria vaginalis*), jungle rice (*Echinochloa colonum*), sedges (*Cyperus* spp.) water grass (*Commelina benthalensis*) etc. are common in direct seeded wet land and transplanted rice. Smith *et al.* (1968) estimated the reduction in rice yield, in association with barnyard grass

Weed density, m^{-2}	0	10.5	50	250
Rice yield; q/ha	51	31	15	7

a. Upland Direct-Seeded Rice

In upland rice, both grass and broadleaved weeds are common. It is normally sown at the beginning of rainy season in most soils which creates most favourable condition for weed germination. Among the weeds, *Echinochloa* is the most competitive weed in dryland rice. In 4 weeks competition, it reduces the rice yield by 40%. Other weeds include; *Cyperus*, *Mimosa*, *Digitaria*, *Cynodon*, *Panicum* etc. and among broad leaf, *Amaranthus*, *Protulaca*, *Phyllanthus* etc.

Weed Control measures

1. Hand weeding and mechanical operation with wheel-hoe are still practiced in a small scale.
2. Propanil at 3 kg/ha as post –emergence is widely used to control barnyard grass at 2 to 3 leaf stage efficiently; and also some sedge and grasses. It is a non-residual contact herbicide which succumbs weed growth at early stage.
3. Spraying of benthiocarb or machete at 2 l/ha as pre-emergence, on surface soil in drilled rice can control both grass and broadleaved weeds.
4. 2,4,5-T or MCPA at 2 l/ha as post-emergence, after one month of seeding rice, is effective to control broadleaved weeds.
5. Other selective herbicides include; nitrofen, oxadiazone, molinate, SWEP, DCPA, flurodifen etc.

b. Direct Seeded Wet-Land Rice

In wet-land rice, aquatic weeds grow abundantly and compete heavily. Major weeds are *Monochoria*, *Cyperus*, *Echinochloa*, *Commelina* etc.

Weed control measures

1. Hand-weeding or use of rotary weeder is a common practice.
2. Molinate 2 kg/ha or a combination of molinate + MCPA (0.75 kg/ha) or molinate + 2,4-D (2 kg/ha) controls most of the weeds efficiently. Nitrofen 3 kg/ha applied before seeding rice, followed by 2,4,5-T (2 kg/ha) spray after 5 weeks can control aquatic weeds.
3. Application of machete granule 20 kg/ha (actual) is most efficient.

Weed control in the nursery

1. In seed-nursery, barnyard grass and sedge emerge fast and compete with rice seedlings which can be controlled efficiently with spray of machete or benthiocarb at 1.0 kg ai/ha in wet soil at 5-7 days after sowing.
2. As per-planting, molinate 2 kg/ha can be incorporated in the soil about 5 cm deep at final preparation of seed bed and just before seed broadcasting.

Weed control in the main field

1. Hand-weeding or rotary weeder like Japanese Hand-hoe is used and 2 to 3 weeding are normally practiced after transplanting rice at 10 to 12 days interval. Hand-weeding consumes 250 to 300 man hours/ ha which is found uneconomical. Moreover, sufficient labour cannot be spared for weeding during peak transplanting season.
2. Machete gr. at 20 kg/ha (actual), if applied 5 to 7 days after transplanting, can give excellent control of most of the weeds. *Banthiocarb* gr. at 20 kg/ha (actual) is also equally effective.
3. A newly introduce herbicide, C-288 in granular form at 30 kg/ha (formulation) is found most suitable to control *Salvinia* under shallow water, but it is slightly toxic to soil and water fauna.
4. Nitrofen (7% conc.), oxadiazon 1.5 kg/ha as post-transplant application may be done. *Molinate* 2 kg/ha maybe incorporated in the soil.
5. 2-4-D or MCPA at 1.5 kg/ha can be sprayed only after full tillering phase i.e. 6 weeks after transplanting. These hormonal weedicides should not be used at early stage of rice.

Control of algae in rice field

Alga problem is very sever in poorly drained transplanted rice. Copper sulphate 15 kg/ha, nitrofen or SWEP granuled 4 kg/ha are important algaecides for rice field. These can be applied 3 weeks after seeding or transplanting.

2. Maize (*Zea mays*) Graminiae

Rainfed maize suffers most due to rapid growth of weds, particularly at its early stage when it is highly sensitive to weed competition. The loss caused in maize from 40 to 70%. Nicto *et al.* (1968) observed that the weed maize competition is most critical upto 45 days of growth.

Yield under varying weed situation

	Infested all season	**Weed-free first 30 days**	**Weed-free all season**
Yield (kg/ha)	380	4440	4780

In Kharif maize, the major weeds include: Acalypha, Euphorbia, Commelina, Celosia, Convolvulus, Digitaria, Amaranthus, Trianthema, Phyllanthus sp. etc.

Weed control measures

1. Inter-cultivation in maize should be done at seeding stage. Hand-hoe, spike-tooth harrow and rotary weeder can be used in rows at its early stage. However, these practices have its own limitations and hence the chemical weed control.
2. Atrazine 1.5 kg/ha as per pre-emergence or early post-emergence has become the standard recommended herbicide in maize and hence widely used. It gives excellent control of established weeds also. It has unique selectivity and has a long range of effectiveness and also long residual effect; but 95% will disappear from soil within 4 months, and hence causes no adverse effect to next rotational crop.
3. EPTC 3 kg/ha plus protectant as soil incorporation can control nut-sedge; while dicamba 1.5 kg/ha can kill Trianthema and Portulaca.
4. Other selective herbicides are: simazine in heavy soil; linuron or propachlor 2 kg/ha in fine tilth.

3. Fodder Maize and Teosinte (*Euchlaena mexicana*)

In fodder maize and teosinte, *Trianthema*, *Elusine* etc. are the common weeds. Since seeds are sown thickly for fodder purposes, manual weeding is inconvenient.

Control measures

1. Atrazine 1.0 kg ai/ha as pre-emergence is most effective.
2. Atrazien in combination with alachlor at 1.25 kg/ha or simazine 1.0 kg ai/ha is also effective.

4. Sorghum (Jowar): (*Sorghum vulgare*) Gramineae

Grain sorghum is less tolerant to pre-emergence sprays than maize. Gupta and Singh (1967) estimated the loss of grains and the effect of weed control.

Yield under varying weed situation

	Infested all season	Weed-free initial 30 days	Weed-free all season
Yield (kg/ha)	850	1790	2080

Weed control measures

1. At early stages of sorghum, weeds can be controlled by use of rotary weeder.
2. Herbicidal wild control in sorghum is somewhat difficult than

maize, though the nature of weed species is almost same in both the crops.

3. As pre-emergence, propazine 1.5 kg ai/ha is widely used to control both grassy and broad-leaved weeds. It has a long residual effect.
4. A combination of propazine + atrazine at 0.5 kg ai/ha each is effective.
5. Fluometuron 1.0 kg/ha or atrazien 1.0 kg/ha as pre-emergence, coupled with one weeding at 45 days after sowing, can give effective weed control.
6. As post-emergence, 2,4-D(amine) and dicamba at 0.5 kg ai/ha can be sprayed in rows, to control broad-leaved weeds like, *Amaranthus*, *Trianthema*, *Digera*, *Phyllanthus*, *Convolvulus* sp. etc.

5. Fodder Jowar and Bajra

In fodder jowar and bajra, most common weeds found are: *Cyperus*, *Cynodon*, *Echinochloa*, *Sorghum halepense*; *Amaranthus*, *Phylanthus*, *Celosia* sp. etc.

Weed control measures

1. Atrazine 1 kg/ha as pre-emergence is very effective.
2. 2,4-D at 1.0 kg/ha as post-emergence may be applied, 15 to 20 days after sowing to control broad-leaved weeds.

6. Pearl millet (Bajra): (*Pennisetum typhoides*) Gramineae

Bajra is grown as rainfed crop in arid and semi-arid region on light texture soils. Moolani and Uppal (1966) observed the effect of weed infestation on bajra yield.

Effect of weed infestation on yield

	Infested all season	Infested fist 50 days	30 days	Weed free all season
Yield (kg/ha)	1330	1790	2560	3120

The weed free condition in bajra can increase the yield more than double than the unweeded area. The major weeds grown in drier region are: *Tribulus terrestirs* (*Puncture vine*) *Cenchrus*, *Tephrosia*, *Indigofera* sp. etc.

Weed control measures

1. Hand-weeding and use of row-weeder are usually practiced twice, 25 and 45 days after planting.

2. 2,4-D 2,4,5-T at 1.5 kg/ha as post-emergence application, 30 days after planting is very effective to control all broad-leaved weeds.
3. Propazine or atrazien at 0.75 kg ai/ha as pre-or early post-emergence is also effective.

7. Finger Millet (Ragl): (*Elusine coracana*) – Gramineae

Weed control measures

1. Mechanical hoeing is normally practiced, though time consuming.
2. Propanil 2 kg/ha as pre-emergence or nitrofen 0.5 kg, neburon 0.5 kg or 2,4-D (Sodium) 0.75 kg/ha gives good control of weeds and thereby increase the yield substantially.

8. Wheat (*Triticum vulgare*) and Barley (*Hordium vulgare*) – Gramineae

Wheat and barley are grown in rabi season as grain crops. Weed competition at early stage causes yield loss 20 to 45%. Major broad-leaved weeds are *Chenopodium*, *Amaranthus*, *Digera*, *Melilotus*, *Trianthema*, *Convolvulus*, *Anagalis*, *Vicia*, *Lathrus*, *Sonchus* sp., and among grassy weeds, *Cyperus*, *Avena*, *Phalaris*, *Digitaria* sp. etc.

Weed control measures

1. Use of clean seed and crop rotation are very effective against many seed species.
2. Good seed-bed preparation and thorough cultivation can bring the weeds down.
3. 2,4-D at 1 kg/ha spray, controls most broad-leaf weeds but this hormonal weedicide should be applied only after 5 weeks of seeding i.e., after tillering.
4. Terbutryne 0.6 kg or nitrofen 2 kg/ha as pre-or post-emergence can control *Phalaris minor* effectively.

B. Weed Control in Pulses Legumes

Weed control is a major problem in legumes, because of slow growth of seedlings and hence most of the fast growing weeds smother pulse crops. Weed competition is very severe during rainy period, particularly at early stages of the legume crop. Irrigated summer and winter pulses are also infested with specific weeds. Weed infestation and smothering take place till 30 to 45 days and hence early weed control is essential. Weed control alone can increase pulse production 30 to 50% however chemical weed control is not very much, advocated because of injury to legume crops.

1. In Kharif pigeonpea, cowpea, soybean, blackgram and green gram fields the common weeds found are: *Trianthema*, *Convolvulus*, *Amaranthus*, *Euphorbea* sp. etc. and grassy weeds like *Cyperus*, *Cynodon*, *Echnochloa* sp.
2. In rabi crops like pea, chickpea (Bengal gram), the major weeds found are: *Chenopodium*, *Melilotus*, *Asphodelius*, *Anagalis*, *Phylaris*, *Avena* sp. etc.

Weed control measures

Preventive measures / mechanical control

1. Legume seeds are usually contaminated with weed seeds and it is very difficult to separate them out; hence the use of clean seed is a must. Thick stand of legume crops also can prevent weed growth at early stages.
2. Mowing and spike tooth harrow operation in row crops are effective methods to control weeds at early stages.

Chemical weed control

Sinbar 3 kg or Sencor 2 kg/ha is very effective to control nutgrass and hence 2-3 months after its application, legume crops can be grown.

1. Pigeonpea (*Cajanas cajan*) – Papilionaceae

As pre-emergence alachlor 2 kg, prometryne 2 kg or ametryne 2 kg/ha give effective weed control.

2. Cowpea (*Vigna sinensis*) – Papilionaceae

1. Prometryne and terbutryne each at 1 kg/ha as pre-emergence application are effective against broad-leaf weeds.
2. Alachlor 2 kg/ha as pre-emergence is very effective in kharif cowpea.
3. Pre-emergence application of trifluratin 1.5 kg, Chloramben 3 kg, Oxdiazon 2 kg or Linuron 0.5 kg ai/ha give adequate weed control in cowpea.

3. Black Gram (*Phaseolus mungo*) and Green Gram (*Phaseolus aureus*) – Papilionaceae

Initial period upto 5 weeks of blackgram and greengram are

sensitive to weeds; which ultimately reflected in reduction of yield as high as 50%. The common weeds found are: *Cyperus*, *Euphorbia*, *Digera*, *Lagasca* sp. etc.

Weed control measures

Conventional method of one weeding at 4 weeks after sowing is effective.

Fluochloralin 1 kg or trifluralin 1 kg/ha as pre-sowing application is quite effective.

Pre-emergence application of alachlor 2 kg or nitrofen 2 kg/ha can give good weed control under rainfed condition in green gram.

4. Broad-Bean (*Dolichos lablab*) – Papilionaceae

1. Pre-emergence application of *chloramben*, *trifluralin*, *molinate*, EPTC at 1 kg/ha each is effective.
2. Dinitramine 0.5 kg/ha as pre-emergence, followed by one weeding at 30 days can control the weeds completely.

5. Cluster Bean (*Cyamopsis tetragonoloba*) – Papilionaceae

1. Pre-planting application of EPTC 2 kg/ha, incorporated in the soil, can control all kinds of weeds.
2. TCA + dalapon at 6.5 kg ai/ha can control Cynodon effectively.

6. Peas (*Pisum sativum*) – Papilionaceae

Peas are mainly infested with broadleaved annual weeds and also some grassy weeds like *Avena fatua*, *Cynodon dactylon*. Mani *et al.*, (1968) calculated to average reduction in yield due to weeds to be about 33%.

Weed control measures

1. A mixture of MCPB and MCPA at 1.5 kg/ha in a ratio of 3:2, as post-emergence spray can control most troublesome weeds, like; Canada thistle (*Cirsium arvense*). MCPB can also control *Spergula arvensis* and *Chenopodium album*.
2. Dinoseb 2 kg/ha can be sprayed early stage when peas are about 15 cm tall, to control difficult weeds like Convolvulus arvensis (bind weed) and also annual grasses for 4 weeks.
3. Pre-planting spray of fluchloralin 3 l/ha and diallate 2 kg/ha are very effective to control many weeds.
4. Pre-emergence application of EPTAM at 1.5 l/ha provides control of nutgrass which is a serious problem. Linuron 1.5 kg/ha as pre-emergence is also effective in peas.

7. Chickpea (Bengal Gram) (*Cicer arietinum*) – Papilionaceae

Chickpea is grown as winter crop. Most common weeds found in chickpea field are: *Chenopodium*, *Melilotus*, *Convolvulus*, *Anagalis*, *Vicia*, *Cyperus*, *Cynodon* etc. Losses in yield is estimated from 40 to 100% according to intensity of weeds.

Weed control measures

1. Two weedings at 30 and 50 days after sowing is the best method of weed control.
2. Preplant application of trfluralin 1 kg, chloramben 2 kg/ha is effective.
3. Atrazine 2 kg, alachlor 2 kg, terbutryne 0.75 kg/ha as pre-emergence can give a good control of weeds including Phalaris minor and other broadleaved weeds.
4. MCPB or 2,4,D-B at 0.75 kg ai/ha spray after 4 weeks of sowing is effective.

C. Weed Control in Oilseed Crops

Among oilseeds, groundnut is the major crop, followed by soybean, linseed, rape and mustard, castor, safflower, sesemum and sunflower. Among these, groundnut suffers most because of its slow vegetative growth and hence smothered by quick growing weeds very fast.

1. Groundnut (Peanut) (*Arachis hypogaea*) – Papilionaceae

Groundnut face sever weed competition at early stage, resulted in as high as 50% yield reduction. Groundnut root grows very fast after planting but the shoot growth is very slow. It also begins to flower early and 30 days after planting, peg begins to enter into the soil. Hence, the crop must be weed-free and loose soil condition upto 45 days which is the critical period for weed competition in rainfed groundnut. The canopy of this close-spaced crop also covers the ground within 45 days and no physical hinderance during its pod-development should be done.

Major weeds in groundnut field include: (*Digitaria*, *Echinochloa*, *Celosia*, *Cleome*, *Cyperus*, *Cynodon*, *Eleusine*, *Convolvulus* sp. etc. It is observed that these weeds remove more quantities of P and K than groundnut. Bhan and Singh (1971) estimated the yield under different management condition.

Comparative effect of hand-weeding and weedicide on yield

	Weedy (check)	Hand-weeded	Prometryne (1 kg ai/ha)
Pod yield (kg/ha)	1870	2780	2930

Weed control measures

1. Two hoeing and weeding 25 and 30 days after sowing during monsoon suppress the weed considerably and can give highest pod yield. This is the most efficient method.
2. Alachlor 2 kg ai, nitrofen 2 kg ai or Promotryne 1.5 kg ai/ha as pre-emergence, can be applied one week after planting so as to reduce the risk of injury the groundnut. This application can be followed by one hand-weeding after 3 weeks to keep the crop weed free.
3. Dinoseb 2 kg/ha as early pre-emergence, just at groundnut emergence is the best.
4. 2,4-D-B at 1 kg/ha as post-emergence can control broad-leaved weeds.

2. Soybean (*Glycine max*) – Papilionaceae

Soybean is grown both as pulse crop and also as oilseed crop. Soybeans are not tolerant to herbicides and hence post-emergence application should not be done. Bhan and Singh (1974) recorded the effect of pre-planting weedicide on yield.

Comparative effect of hand weeding and weedicides on yield

	Weedy (check)	Hand-weeding	Alachlor (Pre-planting (2 kg/ha)
Yield (kg/ha)	1170	2610	2750

Weed control measures

1. Alachlor 2 kg and propachlor 3 kg/ha as pre-emergence are commonly used at early stage to control mostly annual grasses. To control broadleaved weeds also, these herbicides should be combined with dinoseb or linuron.
2. Dinoseb 2 kg/ha as early pre-emergence is best to control Convolvulus (bindweed) and Xanthium (Cocklebur).
3. Pre-planting incorporation of trifluralin 1 kg or nitralin 1 kg/ha gives good result.
4. Basalin 2 1 + alachlor 1 l/ha in combination gives effective control of both grassy and broad leaved weeds.

3. Linseed (*Linum usitatissimum*) – Linaceae

Linseed is a winter season crop, grown mostly under un-irrigated condition in rice fallows. It is usually sown mixed with winter cereals and gram; hence selection of suitable weedicide is difficult; but as a pure crop, herbicidal control is possible.

In linseed, *Chenopodium*, *Anagalis* and a grassy weed, *Avena fatua* are most common. The comparative efficacy of hand-weeding and weedicide application is as follows:

	Weedy (check)	Handweeded	MCPA (0.5 kg ai/ha)
Yield (kg/ha)	870	1140	1280

Weed control measures

1. Hand weeding once or twice at 15 days interval is the usual practice.
2. Trifluralin 1.1 kg/ha during soil preparation, 2 weeks before drilling the seed, is effective.
3. A combination of MCPA (0.25 kg) + MCPB (0.25 kg/ha) in a ratio of 2:3 is most effective, particularly to control Canada thistle (*Cirsium arvense*) and sowthistle (*Sonchus* sp.) which are most troublesome weeds.
4. EPTC at 2 kg/ha as pre-planting and TCA at 5 kg or dinoseb 0.5 kg ai/ha as post-emergence can control grassy weeds effectively.

4. Rape and Mustard (*Brassica* spp.) – Crucifereae

Rape and mustard are early winter crops and mixed cropping with other winter crops is very common. They grow very fast, produce large leaves and quickly shade the inter rows. In *Brassica* fields, most common weeds are: *Chenopodium* and *Cyperus* spp.

Weed control measures

1. One hand-weeding is sufficient if done timely.
2. EPTAM 2 kg or alachlor 2 kg/ha can control both *Chenopodium* and *Cyperus*.
3. Nitrofen 2 kg or Propachlor 4 kg/ha as pre-emergence 3 days after sowing, can control broadleaved weeds; while dicamba kills perennial weeds like *Cirsium arvensis*.
4. Preplanting application of trifluralin 1 kg/ha or post-emergence application of dalapon 4 kg or TCA 5 kg/ha can control most annual weeds.

5. Castor Bean (*Ricinus communis*) Euphorbiaceae and Til (*Sesamum indicum*) – Pedaliaceae

In castor, weed problem is most intricate because of its tall habit and hence hardly any smothering effect. However, sesamum suffers most because of frequent emergence of kharif weeds.

Weed control measures

1. Early sowing and inter-cultivation are normally practiced.
2. Alachlor 1.75 kg/ha or CDAA at 6 kg/ha as pre-emergence or EPTC at 2 kg/ha as pre-planting application is advocated.

6. Safflower (*Carthamas tinctorius*) – Compositae

1. Nitrofen 1.5 1 alachlor 1.5 L or trifluralin 1 kg/ha as pre-emergence is very effective.
2. As pre-planting, EPTC at 2 kg/ha may be incorporated in the soil.

7. Sunflower (*Helianthus annus*) – Compositae

In sunflower, the common weeds are *Cyperus*, *Cynodon* and *Chenopodium* sp.

Weed control measures

1. Early weeding upto 15 days, keeps the crop vigorous and increase the yield.
2. EPTAM at 1.25 kg/ha as preplanting incorporation or alachlor 1.25 kg/ha as pre-emergence application just before sowing is the best to control weeds.

d. Weed Control in Fiber Crops

1. Cotton (*Gossypium* spp.) – Malvaceae

Cotton crop grows very flow in the initial stags and suffers very badly due to sever weed competition especially for light. Weed takes about 8 weeks to reach maximum cover due to its quick growth, while cotton takes at least 16 weeks to cover 90% space. Hence, the competition for light is more severe, particularly to young cotton plants which is reflected to poor yield. Buchanan and Burns (1970) observed the influence of weed competition lass generally for the first 50 days of growing period.

Days of weed-free maintenance

Days	Weedy (check)	10	20	30	40	50	70	Weed free all season
Yield (kg/ha)	300	550	1200	2100	2500	2700	2700	2700

Due to weed competition at early stages, the yield reduction is ranging from 16 to 63 per cent and hence weed control at early stages is very important. Any delay in weed control may not compensate the losses occurred. Further the presence of weeds in cotton makes harvesting more difficult and reduces the spinning quality of fibre. Keeping the field weed free, cotton yield can be increased by 35 to 40%.

The weed flora predominant in cotton are mostly annual grasses and sedges like *Echinochloa; Cyperus*, *Cynodon*, and low percentage of broadleaf weeds include *Amaranthus Commelina*, *Digera*, *Euphorbia*, *Convolvulus arvensis*, *Phyllanthus*, *Trianthema* and trouble some weeds like *Xantheum strumarium* (cockle bar).

Method of Weed Control

Mechanical control

Certain cultural practices like rotation of crops, mulching, clean cultivation are useful. Two to 3 hoeings and weeding in a 30 day old crop are required to control the weeds at early stage.

Chemical control

1. **Pre-planting** (i) Soil incorporation of fluchloralin (Basalin) at 2.0 l/ha as pre-sowing application upto 5 cm deep is very useful to control mostly annual grasses and some broadleaved weeds. (ii) Pre-sowing application of nitraline, trifluralin and dinitraine at 1.0 kg/ha each; alachlor 5.0 kg/ha or diuron 1.5 kg/ha is very effective (iii) For controlling Cynodon, TCA at 5 kg/ha as preplanting treatment is the best.
2. **Pre-emergence:** (i) Pre-emergence application of fluomaturon 3.0 kg/ha or diuron 1.2 kg/ha is most effective. (ii) Trifluralin 0.75 kg/ha and nitralin 3 kg/ha can control most grasses and *Sorghum halepense*. (iii) Nitrofen 1.5 kg/ha as pre-emergence followed by MSMA at 5 l/ha post-emergence can give good weed control. Monuron at 1.5 kg/ha may be used on excessively heavy soil. (iv) Chloroprophan at 6 kg/ha can safely

be used both heavy and light soils as pre-emergence.

3. **Post-emergence:** MSMA and DEMA at 2.0 kg/ha as post-emergence application can control both annual and perennial weeds. However, TCA and dalapon at 1% are well known herbicides to control perennial weeds like Cynodon and Sorghum halepense (Johnson grass).

2. Jute (*Corchorus olitorius* & *C. capsularis*) – Tiliaceae

Jute is a kharif season crop, usually sown after receiving pre-monsoon showers during April-May. During early monsoon, weed seeds also germinate every fast and smother the young jute seedlings, which is reflected in fibre yield reduction to as high as 30 to 55 per cent, depending upon weed density.

The major weed species include: *Echinochloa*, *Cyperus*, *Setaria*, *Panicum sp. cynodon*, *Digitaria*, *Tridax procumbens* etc. Amongst them, *Echinochloa* is the most competitive weed in jute field.

Weed control measures

1. Two to 3 hand weedlings or use of jute rotary weeder are most common to control weeds at early stage; but during the same period, rice nursery preparation and transplanting coincide and hence labour scarcity arises. Hence, chemical control of weeds would be desirable.
2. One hand weeding at early thinning is 12 days after sowing, followed by dalapan spray at 5 kg/ha plus MSMA 2 kg/ha after 3 to 4 weeks of sowing is most effective to control many annual grasses. Dalapon should be spread cautiously with shielded nozzles so that falling of chemical drips on jute tips and foliage is avoided.
3. Kerb or Sirmate at 3 kg/ha can also control annual grasses as per-emergence application.

e. Weed Control in Sugar Crops

1. Sugarcane (*Saccharum officinarum*) – Gramineae

The plant came of the first crop takes about 3 weeks for germination and during this period, it receives two irrigations which are favourable for weed germination. Those weeds become dense with favourable moist soil condition and arrest the growth of emerged sugarcane shoots. Hence weeding at the initial stages is utmost important. Weed-crop competition at early stage in sugarcane causes 50 to 80% loss in millable cane yield. Weed problem continues into its ratoon crop also.

The plant cane, *Cyperus* sp. is a common weed while in ratoon crop, *Cynodon* (*Burmuda grass*) is usually prominent at early stage of sugarcane.

Other major weeds include: *Panicum*, *Andropogen*, *Echinochloa*, *Sorghum halepense* etc. among broad-leaved weeds. *Trianthema*, *Euphorbia*, *Amaranthus*, *Digera*, *Vicia.*, *Ipomoea*, *Xanthium*, *Anagalis*, *Malilotus* etc. are most common.

Weed control measures

1. Clean weeding for first 100 days of crop growth can increase can yield by 45%.
2. After completion of cane germination, one hand weeding and two row-cultivations during summer months are practiced but this method of operation is labour consuming and very costly. When sugarcane stand is poor, the weeds grow abundantly and smother the crop.
3. Simazine or atrazine at 2.0 kg/ha; supplemented with one hoeing + earthing later, control weeds more efficiently.
4. Atrataf 50W (atrazine) at 5 kg/ha + diuron at 1.0 kg/ha, is the economic effective dose.
5. Parquet (1.0 kg/ha), MSMA and DSMA (2 kg/ha) can be spread before 10% emergence of sugarcane to kill young weeds.
6. Mixture of Atrazine (0.25 kg/ha) plus 2,4-D (1.0 kg/ha) controls especially nutsedge.
7. Post emergence Picloram at 2.0 kg/ha is used for control of mostly broadleaved weeds.
8. Established grasses like Sorghum halepense (Johnson grass) and Cynodon can be controlled with dalapon at 1.0% spray.

2. Sugarbeet (*Beta vulgaris*) – Chinopodiaceae

Sugarbeet is a rabi season crop which is frequently irrigated. The weeds germinate fast even before optimum germination and establishment of sugarbeet seedlings and smother the crop. In early planted crop, *Trianthema*, *Portulaca* and *Avena fatua* (wild oat) become more vigorous. Other weeds include *Cyperus*, *Chenopodium album*, *Melilotus*, *Anagalis*, *Convolvulus* sp. etc. Mani *et al.* (1968) observed that sugarbeet suffered most due to weed infestation at early stages and calculated the reduction in yield due to weeds was as high as 70.3 per cent. It was also further observed that weed free maintenance plot recorded the sugarbeet yield substantially.

Days of weed free maintenance after planting of crops

	Weed infested all season	Weed free days	
		10	20
Yield (Q/ha)	100	268	446

Weed control measures

1. Handweeding is the common practice to control weeds in sugarbeets though this practice is costly.
2. Herbicide application is found to be less costly and highly efficient. Nitrofen at 2.0 kg/ha or alachlor at 3 kg ai/ha as per-emergence can control the weeds effectively.
3. Barban at 0.75 kg/ha as post-emergence can control *Avena fatua* at 2 leaf stage.

f. Weed Control in Tobacco (*Nicotiana tabacum*) – Solanaceae

Weeds compete with tobacco both in nursery and in field. Weed species varies widely and some weeds like *Orobanche* sp. (broom-rape) regrow quickly. The weed growth is composed of *Cyperus* sp., *Cynodon dectylon*, *Medicago* sp., *Anagalis* sp., *Euphorbia* sp. etc.

Weed control measures

1. Ploughing in summer gives good control of certain weeds like *Cyperus* sp., Euphorbia sp. etc.
2. In tobacco nurseries, soil incorporation of diphenamid at 3 kg/ha is very useful to control weeds.
3. In the mainfield, EPTC at 3 kg/ha spray is most effective and causes no adverse effect on tobacco. It is very difficult to control lorobanche, because of its regrowth however, swabling with Paraffin gives quick knock down effect, hence is applied at its emergence.
4. Dalapon 5 kg/ha or linuron 3 kg/ha as pre-emergence application is very effective against *Orobanche*.

G. Weed Control in Vegetable Crops

Heavy losses occur due to weeds in vegetables crops. It is estimated that the yield reduction due to weed infestation ranges from 70 to 90% in carrot, 80 to 90% in onion, if the crops are not free from weds first 4 weeks which is the most critical period in affecting crop yields. Hand weeding and hoeing are the usual practice but too expensive and hence more efficient methods of weed control is necessary. The main weed problem in vegetables is the annual weeds which are usually removed by hand. Perennial weeds, if any are removed by digging at the time of harvest or field preparation. Most of the germinating annual weeds can be controlled by Chloroxuron (4.5 kg/ha) or Paraquat + Diquat at 0.75 kg/ha each, or Propham (2.5 kg/ha) or Chloropropham (0.55 kg/ha) + Diuron (0.25 kg/ha), Propachlor at 4.5 kg/ha can also applied in granules for the control of most annual weeds.

1. Potato (*Solanum tuberosum*) – Solanaceae

Potato is cultivated in rabi season (Oct.-Nov) in plain areas and as summer crop (March-April) in northern hills. It takes about 3 weeks for complete germination after receiving at least two irrigations. During the long germination phase, most of the broad-leaved weeds emerge. Mani *et al.* (1968) reported the reduction in yield due to weeds was 20% per cent.

Among broad-leaved weeds, *Chenopodium album*, *Vicia* sp. *Lethyrus aphaca*, *Spergula arvensis*, *Oxalis latifolia*, *Physalis* sp., and among mono-cots, *Cyperus rotundus*, *Poa annua* (blue grass) *Echinochloa* sp. *Setaria* sp. etc. are most common. The perennial weeds like *Cirsium arvense* (Canada thistle) and Convolvulus sp. (bind weed) continue to compete potato plant and becomes troublesome for eradication.

Bansal (1974) observed the role of EPTC in weeding potatoes and recorded successful results.

	Weedy check	**Handyweeding**	**EPTC 30 kg/ha (pre-plant)**
Potato yield (kg/ha)	81.25	98.43	114.06

Weed control measures

1. Weeding is normally done by manual labour at the time of ridging but mostly ignored during the germination phase of the weeds and thus crop suffers badly.
2. Herbicidal control of weeds at early stage is essential. EPTC at 2 kg/ha, TCA 15 kg/ha or dalapon at 6 kg/ha as pre-plant incorporation to soil, can be sprayed on the soil and harrowed one day before planting of sprouted tubers.
3. To control Cynodon, dalapon + paraquat (1 kg/ha) as early post-emergence application can knock down most of the

emerged weeds.

4. MCPA at 0.5 kg/ha as pre-emergence herbicide is useful for short duration potato, planted on ridges.
5. Metobromuron at 1.5 kg/ha as pre-or post-emergence herbicide is sprayed two days before crop emergence to kill emerged weed seedlings. It does not pose any residual toxicity to rotational crops. Barban at 0.7 kg/ha can be spread on weeds at 2 leaf stage.
6. Since potato plants are tolerant to Propanil, the herbicide of 1.0 kg/ha can be used as post-emergence chemical control of weed seedlings as blanket spray.
7. For hulm destruction, diquat 0.75 kg/ha or dinoseb 2 kg/ha may be spread on the crop on maturity. Sulphuric acid 12% concentration at 1100 litre/ha is also used for this purpose.

2. Tomato (*Lycopersicum esculcutum*) – Solanaceae

Tomato is cultivated in both irrigated and rainfed condition. The common rainy season weeds are *Cyperus*, *Cynodon*, *Amaranthus*, *Legasca* etc; while in Winter season the weeds include; *Chenopodium*, *Melilotus*, *Vicia*, *Portulaca*, *Anagalis*, *Elusine indica* etc.

Weed control measures

1. Hand-weeding twice is essential
2. Herbicide application in tomato is found cheaper than mechanical or manual weeding. Metribuzine (Sencor) at 0.5 kg/ha at post-planting can keep the field weed free for a long period.
3. Alachlor 2 kg ai butachlor 2 kg ai or nitrofen 2 l/ha as pre-emergence one week after transplanting, is very effective and has no residual toxicity in the next crop.
4. Trifluralin 1.1 kg/ha in soil incorporation, nitralin 0.5 kg/ha before weed emergence and chloramben 3 kg/ha one week after transplanting of tomato seedlings, are very effective.

3. Brinjal (*Solanum melongena*) – Solanaceae

Weed control measures

1. Fluchloralin 1.0 kg ai/ha spray one week after transplanting as pre-emergent spray in moist soil, is very effective.
2. Alachlor 2.5 kg ai/ha as preplant and paraquat 0.5 kg ai/ha as post-planting spray in the interrows with a protective hood, is very effective.

3. DPCA at 10 kg/ha spray as post-emergent application is also useful.

4. Chilies (*Capsicum annuum*) – Solanaceae

Weed control measures

1. Trifluralin 1 kg/ha in soil incorporation, 3 weeks before transplanting is very useful.
2. Pre-transplanting application of nitralin is helpful.
3. Pre-emergence application of alachlor or nitrofen 2 kg ai/ha followed by one hand-weeding after 30 days, can control all weeds effectively.

5. Bhendi (*Abelmoscus esculentus*) – Malvaceae

Weed control measures

1. Incorporation of trifluralin (Basalin) at 1.5 kg ai/ha before sowing is very effective to control almost all kinds of weeds so as to keep the field weed-free atleast for a month.
2. Alachlor or butachlor 1 kg ai/ha as pre-emergence or fluchloralin 1 kg ai/ha one week before sowing seeds, followed by one weeding is very effective.
3. Prometryne 0.5 kg/ha as pre-emergence + one handweeding after 30 days is economical.

6. Cabbage, Cauliflower (*Brassica oleracea* var. *capitata*) – Cruciferae

Cole crops like cabbage, cauliflower, knolkhol etc. are grown in winter season and hence most of the winter weeds like *Chenopodium*, *Melilotus*, *Convolvulus*, *Anagalis* etc. are common besides perennial weeds like *Cynodon* and *Cyperus* etc.

Weed control measures

1. In cabbage, cauliflower and brassel sprout, pre-sowing application of TCA at 10 kg, nitrofen 2 kg, diallate 1.7 kg/ha can control most of the annual grasses.
2. Oxyfluorfen 0.25 kg ai, Trifluralin or fluchloralin at 1 kg ai/ha can be successfully applied one day before transplanting as pre-emergent spray.
3. Alachlor 1.5 kg ai, nitrofen 3.4 kg or propachlor 4.5 kg/ha as pre-emergence, followed by one hand-weeding is effective in cabbage and knolkhol.
4. Porpachlor 4.5 kg/ah as post-emergence can be applied in

cabbage, cauliflower and Brassels sprout.

5. Tenoran 50 WP at 1 kg ai/ha is a most effective weedicide for cole-crops.

7. Cucurbits: Cucumber (*Cucumis sativus*), Muskmelon (*Cucumis melo*) Watermelon (*Citrullus vulgaris*), Pumpkin (*Cucurbita pepo*) and Squash (*Cucurbita maxima*)

Weed control measures

1. Butachlor 2 kg ai/ha one day after sowing seeds or Chloramben 3 kg/ha as pre-emergence, is effective for direct seeded cucurbits.
2. CDEC at 4 kg/ha as pre-emergence can control annual weeds.
3. DCPA at 6 kg or Dinoseb 4 kg/ha as pre-planting incorporation, 2 weeks before planting, can control weeds effectively in muskmelon.
4. Other pre-emergence herbicides are: alachlor 2.5 kg, liuron 0.5 kg, dichlormate 0.4 kg diuron 1.2 kg/ha for effective weed control in cucurbits.

8. Beans (*Vicia faba*) and Greenpeas (*Pisum sativum*) Papilionaceae

Weed control measures

1. Amiben 4 kg or CDEC 4 kg/ha as pre-emergence is effective on most annual weeds.
2. EPTC at 2 kg/ha or DNBP 6 kg/ha or Propham 3.5 kg/ha can be incorporated in the soil at planting time for effective weed control in beans.
3. Chloroxuron 1.5 kg/ha as pre-emergence one day after weed is effective in cluster beans.
4. Alachlor 2 kg ai/ha as pre-emergence followed by one hand-weeding after 30 days of sowing is best in case of green peas.
5. Post-emergence application of MCPB at 1.5 kg or nitrofen 1 kg ai/ha or dinoseb 2.8 kg/ha is very effective in controlling broad-leaved weeds.

9. Leafy Vegetables: Spinach (*Spinacea oleracea*) Chenopodiaceae; and Lettuce (*Lactuca sativa*) – Cruciferae

Weed control measures

1. Pre-emergence application of CDEC or dinoseb at 2 kg/ha is effective.
2. Other pre-emergence weedicides are: lenacil 1 kg/ha can

control effectively in spinach, Chlopropham 1 kg, Propham 5 kg + diuron 1 kg, monuron 1 kg or trifluralin 1 kg/ha.

10. Onion (*Allium cepa*) and Garlic (*Allium sativum*) – Liliaceae

The bulb yield can be increased significantly, if the field is kept free of weeds for initial 45 days. Mani *et al.* (1968) calculated the loss due to early infestation of weeds was as high as 68%.

Weed control measures

1. Light cultivation at 3 leaf stage of onion or garlic can eliminate the weeds.
2. Fluchloralin 1 kg ai/ha as preplanting followed by one weeding after 30 days is very effective.
3. As pre-sowing treatment, triallate 1.5 kg, alachlor 2 kg, Propachlor 4 kg and as pre-emergence, monuron 1 kg, Nitrofen 2 kg + chloroxuron 1 kg ai/ha, one weed after transplanting, is effective.
4. Oxadizon 1.5 kg or ioxynil 0.7 kg/ha as post-emergence at early stage is very effective.
5. For post-harvest desiccation, dimexon 20 kg/ha can be spread.

H. Weed Control in Root and Tuber Crops

1. Radish (*Raphanus sativus*), Turnip (*Brassica napus*) Cruciferae and Carrot (*Daucus carota*) – Umbelliferae

In radish, turnip and carrot fields, the prominent weeds are *Cyperus*, *Cynodon*, *Poa*, *Chenopodium*, *Amaranthus*, *Anagalis*, *Melilotus*, *Convolvulus* sp. etc.

Weed control measures

1. Conventional method of handweeding is laborious; hence pre-planting incorporation or per-emergence spray is most economical.
2. EPTC at 3 kg/ha as preplanting incorporation, TCA 10 kg, trifluralin 1.1 kg or diallate 1.7 kg/ha; followed by nitrofen 2 kg/ha or prometryne 1.1 kg/ha as pre-emergence at 3 leaf stage, can keep the crop weed-free.
3. Linuron 1 kg/ha as pre-emergence followed by Chloroxuron 1 kg/ha, 2 weeks after emergence, can give a good control of weeds.

2. Sweet Potato (*Ipomoea batatas*) – Convolvulaceae

1. EPTC at 4 kg/ha as post-transplanting as soil incorporation in the inter rows of sweet potato setts, can control weeds effectively.
2. Other suitable herbicides are: chloramben 2 kg, CDAA 4 kg/ha.

3. Cassava (Tapioca): (*Manihot utilissima*) – Euphorbiaceae

Weeds become a problem at initial stage of cassava for about 30 to 45 days and afterwards the canopy spreads all over the area.

Paraquat 1 kg ai/ha can be spread between rows with a protective cover, so as to avoid any drift.

4. *Colocasia* and *Xanthosoma* spp. – Araceae

1. Propanil 3 kg/ha pre-emergence can be applied when weeds 5 cm tall.
2. Ametryne or prometryne 3 kg/ha or trifluralin 1 kg/ha as post-emergence is very useful.

5. Yams (*Dioscorea esculenta*, *D. alata*) – Dioscoreaceae

Atrazine 3 kg + TCA 5 kg/ha as per-emergence is effective.

I. Weed Control in Annual Spices

1. Turmeric (*Curcuma domestica*) and Ginger (*Zingiber officinale*) – Zingiberaceae

These rhizomatous crops are infested with weeds at early stages.

The common weeds are: *Trianthema*, *Euphorbia*, *Phyllanthus*, *Cyperus*, *Echinochloa* etc.

Weed control measures

Application of alachlor 2 kg/ha as pre-emergence can control the weeds efficiently upto 3 to 40 days.

2. Coriander (*Coriandrum sativum*) – Umbelliferae

Triallate 1.7 kg/ha as pre-sowing incorporation in the soil, one day before sowing seeds or atrazine 1.5 kg/ha as pre-emergence application on moist soil is very effective.

Prometryne 1.7 kg/ha as post-emergence at 3 leaf stage of broadleaved weeds is effective.

J. Weed Control in Fodder Legumes

1. Lucerne (Alfalfa): (*Medicago sativa*) – Papilionaceae

Early season weeds germinate along with Lucerne crop in wet seedbeds and compete. Amongst the weeds grown, *Portulaca*, *Cuscuta* sp. are most common.

Weed control measures

1. Use of clean weed-free seeds, rotation of crops and treating seeds in 10% salt solution to eliminate undesirable materials are essential.
2. Close mowing of forage crop can suppress growth of Canada thistle.
3. Preplanting soil incorporation of EPTC 3 kg; and pre-emergence application of diuron 1 kg ai/ha is effective.
4. Dichlobenil 2 kg or chloroprophan 2 kg/ha applied in granular from just after irrigation, can control dodder (Cuscuta sp.) for 6 weeks.
5. Trifluralin 0.75 kg/ha at sprouting of Lucerne can control weeds for a longer period.

2. Berseem (*Trifolium alexandrinum*) – Papilionaceae

Weed problem in berseem is similar to that of Lucerne. Since berseem is sown as broadcast thickly, manual or mechanical weeding is difficult.

Chemical weed control

1. EPTC 1.5 kg/ha as preplanting incorporation in soil followed by post-emergence application of MCPB 0.5 kg/ha can control weeds efficiently.
2. Dinoseb 4 kg ai/ha or Simazine 1 kg ai/ha after the first cut is beneficial.

K. Weed Control in Plantation Crops

Weed control around the basin or in continuous about 2 m strip along/across the tree rows is essential. Mechanical cultivation causes injury to shallow feeding roots and become hinderance by lower fruiting branches. Hence, use of herbicides is desirable but their application should be avoided during fruiting season of fruit crops. In coffee, tea and young rubber plantations, some noxious weeds like *Eupatorium*, *Oxalis*, *Paspalum*, *Polygonum*, *Pennisetum*, *Setaria* and *Cynodon* infest the crop very vigorously.

The cost of manual weeding is estimated about 17% of total cost of annual cultivation; moreover, labour availability in plantations for timely weeding is also a problem and hence chemicall weed control is desirable.

1. Coffee (*Coffea arabica*) – Rubiaceae

1. A mixture of paraquat 0.5 kg/ha plus simazine 2.5 kg/ha spray on young weeds of 15 cm height is the best; but shield nozzole has to be used to avoid drift to young coffee plants.
2. TCA 5 kg or Simazine 2 kg/ha as pre-emergence can control annual weeds.
3. Ametryne 2 kg/ha as post-emergence against young weeds is suitable.

2. Tea (*Camellia sinensis*) – Theaceae

Weeds are a great problem in new tea clearings and pruned fields. Weedicides like dalapon, 2,4-D, MSMA, paraquat, diuron simazine and glypohsate are the recommended herbicides for weed control in tea and amongst them, glyphosate (round up) has become as promising brad spectrum weed killer in recent years. It is reported that crop losses upto 12% due to broadleaf weeds and upto 21% due to grassy weeds competition in tea in India.

Weed control measures

1. A combination of glyphosate 1.2 kg + 2,4-D 1 kg/ha can control both grass and broadleaved weeds.
2. Simazine at 2 ppm or 2 kg/ha as pre-emergence is safe to apply. Ametryne 2 kg/ha can also be applied as pre-emergence.
3. Oxyflurfen 0.25 kg ai/ha as pre-emergence, followed by second spray of xoyfluofen + paraquat 0.25 kg ai/ha or + 2,4-d 0.8 kg/ha can control 20% regeneration of noxious weeks.

3. Rubber (*Hevea brasiliensis*) – Euphorbiaceae

Young rubber plants are mostly infested with Eupatorbium and several grasses.

Weed control measures

1. Growing of cover crops like *Calopogonium* sp., *Mimosa invisa* will cover the interspace and smother noxious weeds.
2. Dalaon 15 kg/ha (actual) or MSMA 2 kg/ha or neburon 2.5 kg/ha can control weeds effectively in young rubber plantations.
3. Fluometuron 2 kg/ha can be applied in established rubber

plantation.

4. Coconut (*Cocos nucifera*) – Arecaceae

Weed control measures

1. Growing of cover crops like *Mimosa invisa*, *Calopogonium mucunoides* in between coconut palms is most practical method of controlling weed growth and is found beneficial.
2. Dowpon 5 kg ai/ha can control *Cyperus rotundus* and grasses most effectively (85% mortality of weeds). DSMA at 3 kg/ha is equally effective to control nut sedge.
3. Bromacil 5 kg ai/ha or 2,4-D at 1 kg ai/ha as post-emergence application, followed by application of dalapon 10 kg/ha can control several weeds upto 8 months.
4. A combination spray of dalapon 10 kg (actual) + 2,4-D 2.5 kg/ha (actual) or 2,4-D + paraquat (1 l/ha each) can control the weeds for 3 to 4 months.

L. Weed Control in Horticultural Plants

In horticultural crops, conventional method of hand-weeding is costly and labour intensive; which can be considerably reduced by adopting chemical weed control.

1. Citrus (*Citrus* spp.) – Rutaceae

1. Periodical cultivation like mowing, mulching, is very effective but difficult.
2. Paraquat 1.5 kg/ha as a post-emergence herbicide can be used in citrus.
3. Monuron or diuron at 2 kg/ha being a citrus tolerant weedicide, can be applied twice in a year, preferably May and October.
4. Bromacil 2 kg ai + diuron 2 kg ai/ha, simazine 2 kg or diuron 2 kg/ha is very effective for weed control in citrus seedlings.
5. EPTC at 1.5 kg/ha as preplanting incorporation can control nutsedge effectively.
6. A combination of diuron 2 kg + bromacil 2 kg/ha can control Cynodon and several other grassy weeds.
7. Dalapon 5 kg/ha at 3 leaf stage as post-emergence can control grassy weeds.

2. Banana (*Musa sapientum*) – Musaceae

Banana roots are mostly spread on surface layer of soil and hence mechanical cultivation is prevented. Till the banana crop canopy is spread fully and covers the entire area, application of weedicides is required to control major weeds.

Weed control measures

1. Alachlor 2 kg/ha or diuron 2 kg/ha as pre-emergence is the best to check weeds most effectively.
2. Paraquat 1.5 + 2,4-D at 3 kg/ha as post-emergence may be spread periodically.
3. Diuron 2 kg ai followed by paraquat 1 kg ai/ha is effective in controlling Eleusine, Setaria and other weeds. Simazine 3 kg/ha is also equally effective to control the weeds.

3. Pineapple (*Ananas sativus*) – Bromeliaceae

The common weeds in pineapple field include; *Cynodon*, *Cyperus*, *Digitaria*, *Amaranthus*, *Euphorbia* etc.

Weed control measures

1. Manual weeding is very difficult because of hinderance of thorny leaves abnormally spread. Hence chemical control is followed.
2. Bromacil 2 kg ai/ + diuron 2 kg ai/ha as pre-emergence is the best to keep the field weed-free for 4-5 months without much injury to the crop.
3. Simazine 3 kg/ha as pre-emergence, 15 days after planting is effective.

4. Grape (*Vitis vinifera*) Vitaceae, Plums (*Prunus Domestica*) – Rosaceae, Apple (*Pyrus malus*) Rosaceae, Gooseberry (*Physalis peruviana*) Solanaceae; and Strawberry (*Fragaria vasca*) Rosaceae

Weed control measures

1. Atrazine or diuron 2 kg ai/ha as pre-emergent spray, followed by paraquat 1 kg /ha can control most of the weeds.
2. Dichlobenil 8 kg, terbail 1.5 kg/ha in soil incorporation can control many weeds.
3. Dalaon 5 kg/ha as foliar spray can control grasses at 3 to 4 leaf stages.
4. EPTAM at 75 l/ha as pre-emergence, followed by dalapon 8.75 kg/ha as post-emergence can control Cyperus rotundus effectively.

M. Weed Control in Ornamental Trees, Shrubs and Flowering Plants

1. Ornamental Trees

Annual weeds can easily be controlled in established gardens both by mechanical and chemical means. Perennial weeds like *Agropyron repens*, *Convolvulus arvensis* are difficult to dig out. Hence herbicides can easily be used around ornamental trees and shrubs.

Weed control measures

1. Chloroxuron 5 kg or paraquat 0.75 kg + diquat 0.75 kg/ha or simazine 1 kg + 2,4-DES at 4 kg/ha can be applied to control mostly annual weeds.
2. Dichlobenil 6 kg/ha or Propachlor 4.5 kg/ha are applied in granular from to control perennial weeds.
3. Lenacil 2 kg/ha as pre-emergence can be incorporated in the soil to control herbacious weeds.

2. Nursery Stock

Weed problem is very serious in flower crops and nursery stock. Application of weedicides may bring the easy solution of this problem. Since flower crops are different in their growth habits, suitable weedicides for different types of crops will be selected.

3. Corms, Bulbous Ornamentals

As a pre-emergence contact herbicides, paraquat at 1.0 kg/ha can be applied.

As a pre-emergence residual herbicide, simazine at 0.8 kg/ha is the best which can be applied immediately after planting corms or bulbous plants. Fenuron at 1.1 kg/ha plus chlorpropham 0.6 kg/ha can control weeds efficiently.

Propachlor at 4 kg/ha is pre-emergence residual herbicide can give good weed control in dahlia.

4. Gladiolous – Iridaceae

1. Alachlor 2 kg ai/ha as per-emergent spray, soon after planting of corms, is effective.
2. Application of Simazine 0.6 kg + lenacil 1.1 kg/ha in moist soil before the crop emergence can control most of the weeds effectively.
3. Chlorpropham at 3 kg/ha may be applied either at pre-crop

emergence or post-crop emergence.
4. Linuron 1 kg/ha spray during kharif season 24 days after corm planting, gives fairly good weed control.

5. Chrysanthemum (*Chrysanthemum morifolium*) – Compositae

1. Pantanochlor at 3.0 kg/ha can be used as Pre-emergence contact application.
2. Trifluralin at 1.1 kg/ha as pre-planting, chloramban at 2.0 kg/ha or chloroxuron 5.0 kg/ha or 2,4-DES at 5.0 kg/ha as post-planting can control weeds effectively.

6. Roses (*Rosa damascene*) – Rosaceae

1. Rose root-stalks can tolerate weedicides well and hence herbicides can easily can easily be used in rose-beds. The suitable herbicides for roses are paraquat 1.1 kg/ha for pre-emergence contact treatment.
2. As pre-emergence residual soil application, lenacil 2.0 kg/ha atrazine or simazine at 1.5 kg/ha and dichlobenil at 6.0 kg/ha are the best for efficient weed control in roses.
3. Simazine 0.8% + 2,4-D 0.05% pre-emergence shows the most promise.

N. Use of Herbicides in Forestry

1. Herbicides for Forest Nurseries

1. Vaporizing oil at 500 l/ha can be used in the seed-beds as pre-emergence application without any harm to seedlings.
2. Simazine at 2.2 kg/ha can be applied as pre- or post-emergence application in the seedbeds or on transplanted lines of the seedling weeds.
3. Paraquat at 0.5 kg/ha with a wetter can be used in seed beds as pre-sowing or pre-emergence application on germinating weeds.

2. Herbicides for Established Forests

1. In undertaking forest areas, normal spraying equipment are not suitable and hence specialized portable battery operated equipments are used for spraying on weeds.
2. Mixture of 2,4,5-T and 2,4-D are applied to control in large areas.
3. To control woody type grasses and herbaceous weeds, pre-planting or post-planting application of dalapon, Paraquat,

dichlobenil, atrazine, glypohsate and pronamide are used. These weedicides are mostly susceptible to troublesome grassy weeds, like *Agrostis* spp., *Poa annua* etc.

4. In case of woody weeds, 2,4,5-T (ester) + 2,4-D(ester) are used by different type of application such as spraying, basal bark spray, on cut stumps, tree injection etc.

O. Choice of Weed Control in Cropping System

Herbicides are often crop specific and hence it is difficult to find herbicides that control a broad spectrum of weeds without damaging to component crops in the intercrop. In an intensive cropping system, herbicide residues from one plant can adversely affect the next or the companion crop.

1. Atrazine and simazine application in maize or sorghum persists in soil longer and affects the germination of the next crop, black gram; whereas peas, gram, lentil grown in rabi season, are not affected.
2. Residues of trifluralin decrease sorghum stands (44%) and depress grain yield (17%); but peas are not affected and it is also non-toxic to maize-groundnut intercropping system.
3. Alachlor + paraquat mixture is phyto-toxic to sorghum, but alachlor alone has no adverse effect on a crop mixture of sorghum + cowpea + lablab. Alachlor + linuron is found promising in maize + beans mixture. Similarly alachlor + linuron is found promising in maize + beans mixture. Similarly, alachlor + nitrofen has the selectivity for maize + cowpea mixture.
4. Prometryme application as pre-emergence weedicide does not affect the component crops in a mixture or sorghum + cowpea.
5. In a crop rotation, use of atrazine in maize, diuron the cotton and amiben in soybean, has no residual harmful effects of yield.
6. In a cropping pattern maize-cauliflower-groundnut, weeds are effectively controlled. Inter-cropping of green gram or cowpea in cotton; intercropping pigeonpea with green or blackgram and inter-cropping sorghum with groundnut or green gram effectively suppress the weed growth when compared to pure crops.

Thus, the harmful residual effect of weedicide calls for a careful selection of suitable tolerant crops in rotational cropping, particularly where intensive agriculture including sequential and simultaneous intercropping is being practiced.

P. Weed Control in Grass Lands

Weeds in grassland is due to faulty management like over-grazing or under grazing, acidity or nutrient deficiencies in the soil, poor drainage condition etc. The most common grassland weeds are: *Circium arvense* (thistle) *Juncus* spp. (rushes) *Rumex acetosella* (dock).

Weed problems in grassland can be trackled in a systematic way (a) prior to sowing in the field (b) at the preparation of seedbed and in the seedbed; (c) in the newly sown grass or grass-legume mixture during established and (d) in the mature crop.

1. The grassland site has to be prepared by ploughing and repeated cultivation, followed by stubble cleaning by spraying of paraquat (1 kg/ha) or dalapon sodium (6 kg/ha) or glyphosate at 1.5 kg/ha for chemical weed destruction in prior to seedling. Dalapon 10 kg/ha is very effective in renovation of grasslands.
2. In the seedbed also, the seedling weeds that emerge before the grass can be controlled by pre-emergence application of paraquat (0.6 kg/ha) or glyphosate at 1.5 kg/ha.

Ioxynil 0.42 kg/ha + bromoxynil 0.42 kg/ha spray can efficiently control most of the broadleaf weeds.

1. Controlling weeds during establishment of grassland, selective weedicides like MCPA (1.4 kg/ha), 2,4-D (0.5 kg/ha), Mecoprop or Dichlorprop (2.0 kg/ha) or mixture of dicamba with mecoprop of MCPA is sprayed when grown is at 2 to 3 leaf stage.
   For seedling weed control in establishing grass-legume mixture, MCPB 2.2 kg/ha + 2,4-D B 2.2 kg/ha or dinoseb 2.2 kg/ha may be applied.
2. Controlling weeds in established mature grasses and legumes, either MCPA 1.6 kg/ha or MCPB 2.2 kg/ha or 2,4-D B at 2.0 kg/ha or dinoseb 2.2 kg/ha is used to control broadleaf weeds, but to control unwanted grassy weeds, dalapon sodium at 2.0 kg/ha can be applied.

Q. Control of Weeds in Lawns and Paths

1. In Newly Sown Lawns and Established Lawns

Control of *Cyperus rotundus* and dicot weeds form and integrated part of proper maintenance of lawns.

1. In newly sown lawns, ioxynil at 1.1 kg/ha can be applied by watering can at seedling stage of the lawn grass, to control broadleaved weeds.
2. In the established lawn, both cultural method and chemical weed control can be practiced. Those lawns which are ever neglected, need renovation by replacing with fresh actively growing grass, mowing, fertilization and good management.

In addition, broadleaf weeds can easily be controlled by application of weedicides such as 2,4-D amine or 2,4-D sodium (1.5 kg/ha) dichlorprop 2.2 kg + 2,4-D 1.5 kg/ha or fenoprop 1.5 kg with 2,4-D or mecoprop 2.0 kg/ha with 2,4-D spray at seedling stage. Fernoxone (2,4-D sodium) is absorbed through leaves and transplanted to the roots of dicot seeds and sedges, without affecting standing lawn grass *Cynodon dactylon*.

2. Moss Control in Lawns

Moss invades lawns due to poor drainage, close mowing, heavy shade, very acid soil conditions, nutrient starvation, drought condition, disease or pest damage of lawns.

1. The determination in health of the turf can be improved by proper management live improvement of drainage, liming over acid soils, correct fertilization, reduction of shade, in good aeration, avoiding any compaction of soil etc.
2. In addition, lawn sand mixture (3 parts ammon sulph + 1 part ferrous sulphate + 10 parts of sand as career) can be applied at the rate of 140 gm per metre square area in early morning.
3. Mercurous chloride (Calomel) at 1 gm/m^2 can be applied, while on hard surfaces on paths etc., chloroxuron at 10 kg/ha can be applied by watering.

3. Weed Control on Paths and Hard Surfaces (which are never dug or disturbed)

1. Handweeding is not practical and economical. Hence herbicide application is the only practical solution to solve the weed problem in such areas.
2. Dalapon (sodium) at 25 kg/ha or paraquat (0.75 kg/ha) + diquat (0.75 kg/ha) or Simazine (3.5 kg/ha) with amino-triazole (.17 kg/ha) or Sodium chlorate at 100 kg/ha may be applied by watering can to control all annual weeds and many established perennials.

3. Dichlorbenil at 15 kg/ha as granules is the best for the control of all annuals and perennials.

R. Weed Control in Non-Cropped Areas

Non-selective weed control means indiscriminate destruction or suppression of existing vegetation for a considerable time and to maintain ground free of all growth for a protracted period. The existing weeds are completely killed both above and below ground.

1. Control of Herbaceous Vegetation in Non-Agricultural Land

Herbaceous vegetation in the non-agricultural land include roadsides, railway embankments, canal and river banks, cemeteries, military campus, grassed areas on airfield and golf course areas, housing estates etc.

In non-crop land areas, tall-growing unwanted growth of ground cover is suppressed and low-growing grasses are encouraged.

1. To prevent establishment of scrub, mowing, (or burning) or high power rotary mower is used of vegetation management to some extent but herbicidal control is more convenient and desirable.
2. In bare lands, sterilants are used.
3. On non-crop land, non-selective herbicide kills the existing vegetation, with or without leaving phytotoxic residues.

2. Herbicides for Non-Cropped Land

1. In railway tract, roadside verges, motorways, adjacent air fields, sodium chlorate at 220 kg/ha or amino-triazole at 6 kg/ha are foliage and root-absorbed herbicides which control both perennial grasses and broad leaf weeds. It damages the top growth and effects in preventing regeneration of deep rooted broad leaf weeds and perennial grasses.
2. A mixture of dalapon (1%) and 2,4-D (1%) is widely used to control grasses, sedges and broadleaf weeds. Wetting agent is also added in the mixture to improve its herbicidal effect and long persistent (5-6 weeks) in the soil. Two simultaneous applications at 10 days interval are required for satisfactory vegetation control.
3. To destroy deep-rooted broadleaved perennial weeds, picloram at 4 kg/ha spray is the best, equally splitted at 10 days interval.

It is a foliage and root-absorbed herbicide. It will keep the field weed free more than a season of *Cirsium*, *Convolvulus*, *Polygonum* and *Sanchus* sp.

4. Glyphosate at 3 kg/ha also controls most of the perennial weeds. Repeated application is needed against a broad spectrum of difficult weeds. It is a foliage absorb herbicide, effects on top growth, prevents regeneration of perennial grasses and broad leaf weeds including *Agropyroan* sp. *Agrostis* sp. *Lolium* sp. *Poa annua*, *Cirsium arvense*, *Convolvulus* sp., *Polygonum* sp. *Rumes* sp. and *Sonchus* sp.
5. Asulum at 7 kg/ha is very effective to control Sorghum halepense (Johnson grass) for 2 to 3 months.
6. Paraquat (0.2%) mixed with amitrole-T (1%) can be used on irrigation channels, canal banks, low lying areas. Prometryme 6 kg/ha, PCP and 2,4-D can destroy both perennial grasses and broad leaved weeds.
7. Petroleum oils with 25% aromatic content can be used as non-selective herbicides.
8. In industrial areas, urban areas and military campus, application of tafazien 50 kg or atrataf at 20 kg/ha dissolved in 1100 L water will keep the area free from weeds for a long time. Prior to such spraying, on through ploughing is helpful, since the chemical can not act on mature plants.
9. Simazine (17 kg/ha) and atrazine (17 kg/ha), bromacil (6 kg/ha) monurno or diuron (13 kg/ha) act as root absorbed residual herbicides and affect on top growth and prevent regeneration of perennial weed, particularly *Agropyron repens*, *Poa annua*, *Cirsium arvense*, *Sonchus arvensis*, *Agrostis* sp and *Lolium* spp.

Recommended mixture for total weed control are

1. Atrazine mixed with 2,4-D or 2,4-D + aminotriazole or sodium chlorate or MCPA.
2. Bromacil mixed with borate or 2,4-D or diuron + aminotriazone or picloram.
3. Diuron _ amminotriazole + 2,4-D or diuron + aminotriazole aminotriazole + dichlorprop + MCPA or sodium + dalapon MCPA.
4. Monuron + sodium chlorate or sodium chlorate + 2,4-D.
5. Simazine + aminotriazole.

Chemical control of woody weeds

Brush and woody species are controlled by: (1) cutting (2) burning (3) bull-dozing (4) rolling / mowing (5) applying chemicals (6) biological control by insects and (7) managed by constant grazing with goats.

Toxic herbicides with heavy doses are used to kill woody weeds. Earlier period sodium chlorate at 200 kg/ha weed used; but recently phenoxy acetic and like 2,4-D, 2,4,5-T etc. for effective control.

The bushy growth of Lantana sp., Cassia sp. small trees of *Acacia Zizyphus*, *Ficus* etc. is required to be checked by different types of applications.

1. Foliar spray is the most effective method for small shrub like Lantana sp., Cassia sp. A complete coverage given using high volume spray so that the plant is completely drenched.
2. 2,4-5-T (ester) compounds are very toxic and very effective in killing those shrubs without regeneration, 2,4-D (ester) + 2,4,5-T (ester) mixed in 1:3 ratio with kerosene or diesel oil are used as spray in foliage and drenched.

Cut stump treatment

1. Those woody shrubs are cut at or near the ground level are treated with suitable herbicide within 4 weeks to prevent regrowth.
2. 2,4-5-T (ester) compounds can be used as paste on the top of the cut stump or brush or as spray. The treated portion should be covered with mud or cowdung. 2,4-5-T ester at 1 litre/200 1 water seem effectively control mixed scrub.

3. Basal Dark Spraying

Basal dark spraying is another method implied to large trees which are treated by brushing the chemicals with a hand lance on the bark to destroy the individual tree.

Bromacil at 5.2 kg ai/ha dry granules or 3% spray in water can be used in dark treatment as brush killer.

Dicamba at 1 litre/2 1 water can be injected 1 ml/cut for effectively killing.

4. Scrub Control

Scrub is the woody growth which have no economic value. Scrub is mostly found in roadsides, railway embankments, ditches and also in orchards.

1. Scrub species can be controlled by mixture of 2,4-5-T and 2,4-D (ester) at 4.0 kg/ha or in a mixture of water and diesel oil 9:1 at 4.5 kg/ha.
2. Picloram + 2,4-D at 2 kg/ha or glyphosate at 2.2 kg/ha as foliar spray can control scrub species effectively.

5. Sterilants Used in Non-Cropped Land

Sodium chlorate 600 kg/ha, Boron 2000 kg/ha, monuron or diuron 20 kg/ha. Simazine 5 kg/ha, bromacil and terbacil 10 kg/ha, metribuzin, 5 kg/ha, TCA 60 kg/ha or fenac 1 kg/ha are used as soil sterilants.

6. Soil Fumigants Used in Non-cropped Land

Mehane (SMDC) at 200 to 400 kg/ha is incorporated in the soil 7 to 10 cm deep by disking and irrigated which is volatized in ordinary temperature and yield phytotoxic vapours in the soil which is effective against weeds from seeds.

CHAPTER 8

Herbicides

Herbicides are chemical compounds which are used to kill or reduce the growth of weeds. For the control of weeds, the first chemicals used were inorganic salts such as copper sulphate, ferrous sulphate, sodium chlorate and sulphuric acid. Even though these chemicals are somewhat crude, they demonstrated the potential of herbicides at a time when no modern herbicides were available.

The fine chemical industries later emerged in early part of 20th century who relied no research to synthesize new chemicals. MCPA was first available as a powder in 1941 when no field sprayers were refined and the revolution in herbicidal weed control practice began with the discovery of selective herbicides, MCPA and 2,4-D in 1941-42. They were found cheap, simple to manufacture, easy to apply and requiring less water than inorganic salts.

Initially salt and ester formulations of herbicides were produced which include: MCPA (1941), 2,4-D (1942), 2,4,5-T (1944), propham (1948), 2,3,6-TBA (1952), dinoseb (1954), dicamba (1965) for selective weed control in cereals.

For grassy weed control in broad-leaved crops, propham a carbamate herbicide was first introduced in 1951, followed by other carbmates and thio-carbamates, viz. Chlorpropham (1951), barban (1958), diallate (1960), triallate (1961), carbetamide (1966), phenmedipham (1967) and asulum (1968).

Subsequently, new groups like substituted ureas and triazines were produced in 1950s, and 1960s for the control of grassy weeds in both broadleaved and cereal crops. The first substituted urea herbicide was monuron (1951), followed by fenuron (1951), diuron (1951), meburon (1951), monolinuron (1900), chloroxuron (1960), linuron (1960), chlorbromuron (1961), metobromuron (1963), metoxuron (1968), isoproturon (1971) etc.

Triazines are an outstanding diverse group of herbicides, from soil-acting residual total herbicides to contact foliar-acting herbicides. The triazines may be divided into three sub-groups (i) chlorotriazine (simazine, 1956, propazine 1957, atrazine 1958, cynazine 1966); (ii) methylthio-triazine (ametryne-1960, simetryne-1962, desmetryne-1964, terbutryne-1966); and (iii) methoxy-triazine (prometon-1960).

Over the past 45 years, more than 10,000 herbicides have been screened, out of which about 1,000 compounds must have reached the trial stage and as a result, about 200 herbicides are at present in commercial manufacture. M/s. Shell International Chemical Company, California reviewed the pesticide market at end-user level for the world in 1978 (monitary values in $ million).

Pesticides	**1978 ($ million)**	**%**
Herbicides	4068	45
Insecticides	3052	34
Fungicides	1522	17
Others	380	4
Total	9022	100

Thus the progressive development of effective weed killers have now become an integral part of production process in developed and developing countries and as a result, methods of growing all crops and crop production methods have practically been changed. The acceptance of cereal herbicides was rapid and by the late 1950s about 65% of the crop was sprayed annually. The use of herbicides further accelerated and become widespread in late 1960s in most of the crops grown.

When herbicides are introduced in a crop production system, the soil management methods have been modified, frequent soil cultivation is discontinued, soil properties significantly changed, less nutrients are used, better use of natural and supplementary moisture by crops, higher crop-root densities and greater exploitation of fertile surface soil by cultivated crops.

Herbicides – Classification and Selectivity

The herbicides are classified into two: (a) Selective and (b) non-selective. On the basis of their mode of action, they are divided into contact and translocated and according to their mechanism of action and whether they are applied in the soil or as foliar spray. Non-selective herbicides are used as soil sterilant or fumigant and also used for aquatic weed control. On the Physiological basis, they may be further classified according to their mode of action and method of application like: (i) Pre-planting, (ii) Pre-emergence and (iii) Post-emergence.

A. Selective Herbicides

Selective herbicides are those which affect only certain weeds, leaving certain crops unharmed, like pre-emergence application of atrazine in maize crop. But the selectivity depends on the amount of chemical applied and the way they are used. The selective weedicides are used in agricultural and horticultural fields. 2,4-D, alachlor, atrazine etc. are selective herbicides used in mixed population.

B. Non-selective Herbicides

Non-selective herbicides like paraquat, are used to control a wide range of vegetation indiscriminately because they are toxic to all plants and highly susceptible to living plant tissues. These are for non-selective use on waste lands, rail tracts, roadsides, industrial sites and aquatic weed control.

Soil Application

1. Herbicides are spread on the soil surface or incorporated at pre-emergence or post-emergence stage of the crop. In moist soil, the herbicidal action is prominent; however, selectivity depends on the amount applied and the way they are used. The weedicides used are: simazine, atrazine, molinate etc.
2. Non-selective herbicides when used in the soil, act as soil sterilant. Herbicides of this group prevent all plant growth by mass killing for a longer period when present in the soil and the soil remains sterile. For this purpose, a high dose is recommended. Those herbicides include: Bromacil, TCA, Diuron etc.
3. A wide range of soil fumigants are used for mass killing and to keep the areas free of plant growth. Fumigants form as a vapour or gas and diffuses through the soil and thus they are used to kill all weed growth before starting a new plantation. The soil fumigants used are carbondisulphide, cynamid, methyl bromide etc.

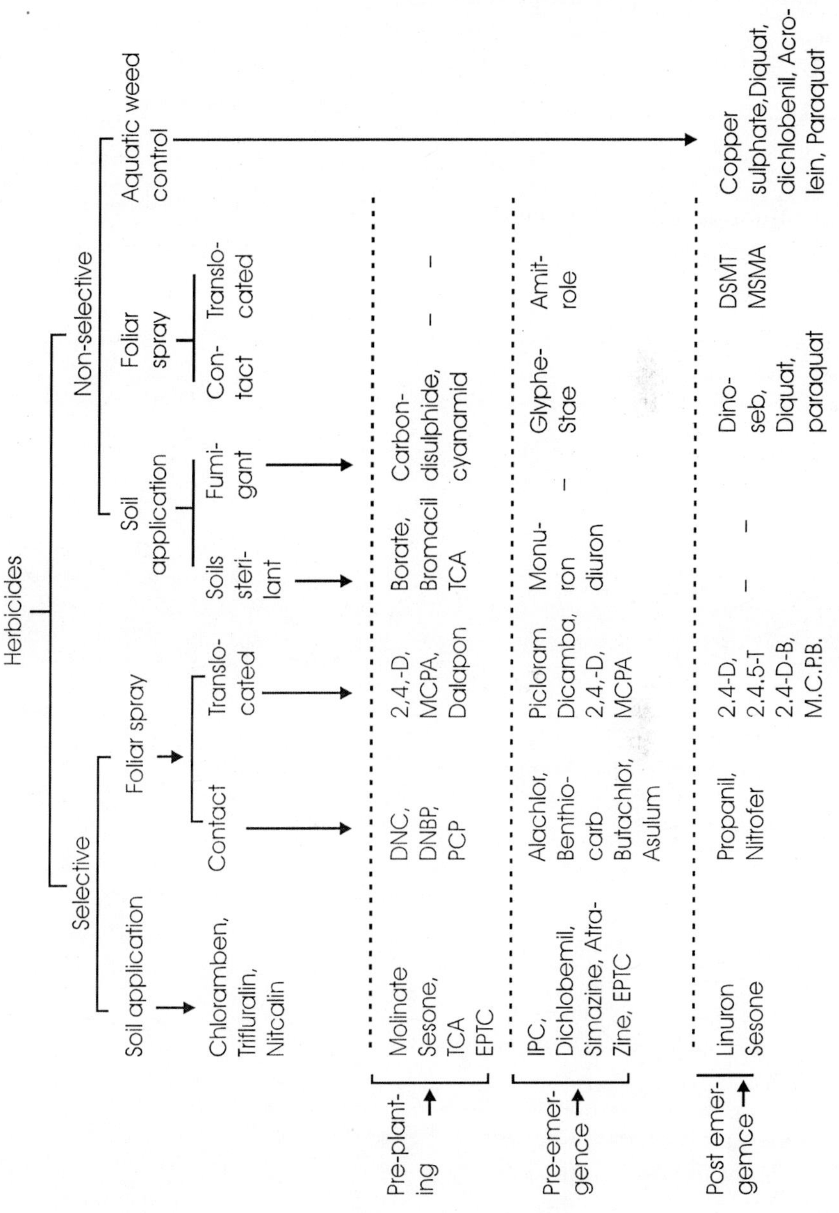

Herbicides
Selective
Non-selective
Soil application
Foliar spray
Soil application
Foliar spray
Aquatic weed control
Chloramben, Trifluralin, Nitcalin
Contact
Translocated
Soils sterilant
Fumigant
Contact
Translocated
Pre-planting
Molinate Sesone, TCA EPTC
DNC, DNBP, PCP
2,4,-D, MCPA, Dalapon
Borate, Bromacil TCA
Carbon-disulphide, cyanamid
–
–
Pre-emergence
IPC, Dichlobemil, Simazine, Atrazine, EPTC
Alachlor, Benthiocarb Butachlor, Asulum
Picloram Dicamba, 2,4,-D, MCPA
Monuron diuron
–
Glyphestae
Amitrole
Post emergemce
Linuron Sesone
Propanil, Nitrofer
2.4-D, 2.4.5-T 2.4-D-B, M.C.P.B.
–
–
Dinoseb, Diquat, paraquat
DSMT MSMA
Copper sulphate, Diquat, dichlobenil, Acrolein, Paraquat

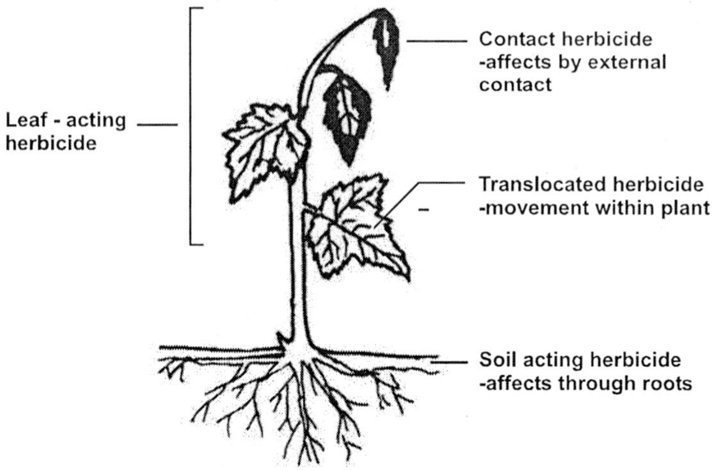

Fig. 1: Illustration of contact, translocated and soil applied herbicides

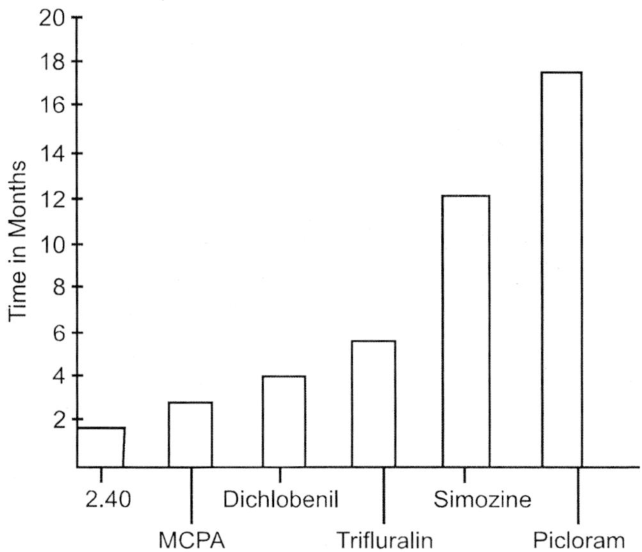

Persistence in soils of selected herbicides

Foliar Application

The action of herbicides as foliar spray depends upon rate and time of its application and its judicial use. These herbicides can be placed in two categories: viz. (i) contact and (ii) translocated, according to their action on a plant. These types vary in their composition, absorption ability, selectively to species and translocation behaviour.

i. Contact Herbicides

These herbicides kill the part of the plant with which they come in contact by foliar spray as they are directly toxic to living cells, such as phloem. Hence, the distribution of spray chemicals should be uniform for efficient weed control by contact herbicides; like alachlor, butachlor, propanil, nitrofen, paraquat. These are mostly effective against annual weeds. In case of perennial weeds, repeated application may be necessary and the weedicides like glyphosate, diquat may be used. A wetting agent is sometimes added to encourage its spread to those parts that have to be destroyed.

ii. Translocated Herbicides

1. These are systemic herbicides which are readily absorbed by either the roots (in case of soil application) or above ground parts and are translocated through the plant system and thus become effective, even in killing the roots of perennial weeds. Growth regulators like 2,4-D, 2,4,5-T, MCPA, Picloram etc. are called translocated herbicides which are usually effective on certain plants (broad-leaved) but no on others and hence can be used selectively, killing certain plants without injuring the cultivated crops. These chemicals enter living plant tissue through leaves or roots or both root and shoot and their effect is to upset some physiological process.
2. Non-selective translocated herbicides are very few (e.g. DSMA, MSMA, Amitrole) and hence repeated application may be needed to kill perennial weeds. The susceptibility can vary with the age of weed, at a stage in growth and hence wetting agents like Tepol, miller-aide may increase efficiency.

Aquatic Application of Weedicides

A few chemicals are sued for control of aquatic weeds as spray on floating weeds like water hyacinth, *Salvinia*, *Pistia* or by dissolving them in water on submerged weeds like *hydrilla*, algae, *Chara*, *Najas* etc. The weedicides used are: copper sulphate C-288, dichlobenil, acrolin etc.

Methods of Herbicide Application

The time and method of chemicals application depend upon the type of crop, type of weed, condition of soil and its environmental factors. Considering all these factors, suitable method of herbicides application have been evolved for satisfactory weed control with minimum loss to the cultivated crop. In some crops, combination of manual or mechanical-cum-herbicidal weed control is essential. The method of herbicide application may be broadly classified into four: (i) Pre-planting (ii) Post-planting (iii) Pre-emergence (iv) Post-emergence.

i. Pre-Planting/Pre-Sowing

In pre-planting treatment, the herbicides are applied in seed-bed or in the main field, incorporating into the soil, usually 20-30 days prior to planting or transplanting in the main field, so as to kill most of the weed seeds.

ii. Post-Planting

In post-planting method, the herbicides are sprayed after planting the crop.

iii. Pre-Emergence

Pre-emergence treatment is made prior to the emergence of specific crop. Mostly contact herbicides are used in this method. The weedicides are applied after the weeds have emerged, before the crop emergence and used as efficient contact herbicides that do not persist in toxic form in the soil. It is to be ensured that the plot remains weed-free when the crop emerges and thus keeps the field free from weeds in early stags of the crop which is mostly needed.

iv. Post-Emergence

Post-emergence treatment is given after emergence of a specific crop or weed; especially post-emergence of the crop.

Translocated herbicides are less influenced and they usually interferes with the normal development of susceptible cultivated plants and several weeks may elapse between application of herbicide and death. Hence the rate, drop size and uniformity of application have to be determined prior to foliage application.

The residual toxicity of heavy dose of herbicidal application to control certain perennial weeds like nutgrass, can be avoided by allowing sufficient time interval for planting the crop.

Herbicides – A Chemical Classification

The chemical classification of herbicides brings into group of all compounds having specific characterization. These chemical affinities are needed to tackle the weed problems.

Most of the herbicides are known by 'common name' and 'trade name', the trade name is usually referred to in its formulation offered by the companies. Their molecular structure and formulation are given by their structural formulae. Many of them are homologous and analogous of a basic structure. The chemical classification of different groups of herbicides, both inorganic and organic compounds, was done by Brian and Crafts (1961) as under. The organic herbicides are divided into families where a chemical group in common to a number of herbicides.

Inorganic herbicides

	Chemicals	**Structural formula**	**Mol. wt.**
Acids	Sulhpuric acid	H_2SO_4	98
	Arsenic acid	$H_3ASO_4H_2O$	160
Salts			**58**
Chlorate	Sodium chlorate	$NaClO_3$	58
Borate	Borax, Decahydrate	$Na_2B_4O_7$	201
	Sodium metaborate	$CuSO_4\ 5H_2O$	132
Sulphamate	Copper sulphate	$CuSO_4\ 5H_2O$	178
	Ferric sulphate	$Fe_2SO_4\ 7H_2O$	208
AMS	Ammonium sulfamate	$(NH_4)NH_2SO_3$	114
Nitrate	Sodium nitrate	$Na\ NO_3$	85

Organic herbicidal compounds

Groups/Common name	**Trade name**	**Structural formula**
Aliphatics (Halo-aliphatic acids)		
Dalapon	Dowpon, Tafapon	2,2-dichlor prpionic acid
TCA	Varitox	Trichloro acetic acid
(a) Anilides and Amides		
Alachlor	Lasso	2-chloro-2, 6-diethyvl-N (methoxymethyl) acetanilide
Butachlor	Machete	N-(butoxymethyl)-2-chloro-2', 6' ethyl acetanilide
CDAA (allidochlor)	Randox	N,N-dially-2-chloro acetamide

Groups/Common name	Trade name	Structural formula
Propachlor	Ramrod	2-chloro-N-isopropyl acetanilide
Propanil	Stam F-34, Rogue	3,4-dichloro propionanilide
Pronamide	Kerb	3,5-dichloro (N-1, 1-dimethyl-2-propynyl) benzamide
Naptalam	Alanap	N-1-naphthyphthalamic acid
(b) Acryldehyde		
Acrolein	Aqualin	
Anilines & Nitro-phenols		
Dinitramine	Cobex	2,6-dinitro-3, amino-4-trifluoromethyl-NN diethylaneline
Nitralin	Planavin	4-(Methyl-sulphonyl)-2, 6-dinitro NN-dipropyaniline
Penexalin	Stamp	N(ethylpropyl)-3, 4-dimethyl-2,6-dinitrobenzamine
Trifluralin	Treflan	2,6-dinitro-NN-dipropyl-4-trifluoron methylaniline
Fluchloralin	Basalin	N-propyl-N-(2' chloroethyl)-2,6-dinitro-n-trifluoroin-ethyl-aniline
DNOC	Sinox	2-methyl-4, 6-dinitrophenol
Dinoseb	Dow-weed killer, pre-emerge	4,6-dinitro-2-5-butyl phenol
PCP	Durotox	Pentachlorophenol
Nitrofen	TOX E-25	2,4-dichlorophenyl-p-nitro-phenyl ether
Fluorodifen	Soyex preforen	2,4-dinitro-4-trifluoromethyl-diphenyl either
Arsenicals		
(Organo-arsenic compounds)		
DSMA	Ansar 184, chipco crab kleen	Disodium methyl arsenate
MAA	Ansar, Weed-E-Rad	Methanearsenic acid
MSMA	Ansar-529, Daconate	Mono-sodium methyl arsonate
Benzoies & Phenyl acetic acid		
Chloramben	Amiben, vegiben	3-amino-2, 5-dichlorobenzoic acid
Dicamba	Banvel-D	2-methoxy-3, 6-dichlorobenzoic acid
Fenac	Fenac	2,3,6-trichlorophenyl acetic acid
Oxyfluorfen	Goal	2-chloro-1(3-ethoxy-4-nitro

Groups/Common name	Trade name	Structural formula
		phenoxy)-4 (trifluormethyl) benzene
Carbamates, Thio-carbamate & (b) carbonilate		
(a) Alkyl		
Dichlormate	Sirmate	3,4-dichlorobenzyl methyl carbamate
Aryl		
Asulam	Asulox	Methyl sulfanilyl carbonate
Barban	Carbine	4-chloro-2-butynyl m-chloro-carbanilate
Chlorpropham	CIPC	Isopropyl m-chloro carbanilate
Propham	IPC	Isopropyl carbanilate
Thio-carbamate		
Benthiocarb	Saturn, bolero	S-(4-chlorobenzyl)N, N-diethyl thio-carbamate
EPTC	EPTAM	S-ethyl dipropyl thiocarbamate
Diallate	Avadex	S-(2,3-dichloro allyl) di-sio-propyl-thio carbamate
Triallate	Avadex BW	S-(2,3,3-trichloro allyl)di-iso-propyl thio carbamate
CDEC	Vegadex	2-chloro allyl diethyl dithio-carbamate.
Mollinate	Ordram-72	S-ethyl hexahydro-I H-azepine 1-carbothionate
Bentazon	Besagran BAS-3510	3-isopropyl-2, 1,3-benzothiadia zon-4-one-2, 2-dioxide
Carbonilate		
SWEP		Methyl 3,4-dichloro carbonilate
Glycine Derivative		
Glyphosate	Non-0573	N(Phosphono methyl) Glycine
Heterocyclic Compounds:		
Bipyridylium		
Diquat	Reglone	1, 1-ethylene-2, 2 bipyridylium dibromide
Paraquat	Gramoxone	1, 1-dimethyl-4, 4-bipyridylium dibromide
Pyridines		
Picloram	TORDON	4-amino-3, 5,6-trichloro picolinic acid

Groups/Common name	Trade name	Structural formula
Pyridazines		
Maleic hydrazine (MH)	Retard	1,2-dihydo pyridazine 3,6-dione
Pyrimidines uracils		
Bromacil	Hyvar-x, Urox-H	5-bromo-6-methyl-3, -butyl uracil
Terbacil	Sinbar	5-chloro-6-methyl 3-T butyl uracil
Triazines		
Atrazine	Atrataf, residox, gesaprin	2-chloro-4-(ethyl amino)-6-(isopropyl amino)-S-triazine
Simazine	Gesatop, Prinlep	2,chloro-4, 6-bis (ethyl amino)-S-triazine
Propazine	Gesamil, milogard	2-chloro-4, 6-bis (isopropyl amino)-S-triazine
Ametryne	Gesapax	2-(ethylamino)-4 (isopropyl-amino)-6-(methylthio)-S-triazine
Prometone	Pramitol	2,4-bis (isopropyl amino)-6-methoxy-s-tirazine
Prometryne	Gesagard caprol	2,4-bis (isopropyl amino)-6-(methylthio-s-triazine).
Terbutryne	Igran	2(tert-butylamino)-4(ethyl-amino)-6-(methylthio)-s-triazine
Metribuzin	Sencor, Lexone Bey-94337	4-amino-6-tert-butyl 3-(methyl-thio)-as-triazine-5 (4H) one.
Misc, hetero-cyclic Compounds		
Aminotriazole		
Amitrole	Weedazol, Amizol	3, amino-1, 2,4-triazole
Endothall	Endothall	3, 6-endoxohaxa hydrophthalic acid
Oxadiazon	Ronstar	2-butyl-4-(2,4-dichloro-5-isopropyl oxyphenol)-1,3,4-oxadiazolin 5 one
Pyrazone	Pyramin	5-amno-4-chloro-2-phenyl 3 (2H) pyridazinone
Hormone type Herbicides		
Phenoxyl acetic acids		
2-4-D	Ethyl ester weedone	(2,4-Dichlor phenoxy) acetic acid
	Sodium-Fernoxone, Weedox, Tafacide;	
	Amine-Dedweed-40	
	Special formulation – Decamine	
2,4,5-T	Brush killer	(2,4,5-trichloro phenoxy) acetic

Groups/Common name	Trade name	Structural formula
		acid.
MCPA	Chiptox	(4-chloro-2-methyl phenoxy) acetic acid.
∝-phenoxy Propionic acid		
Dichlorprop	2,4-DP	∝-(2,4-dichlorophenoxy) propionic acid
Silves (Fenoprop)	Weedone, 2,4,5-TB Kuron, Fenoprop	2 (2,4,5-trichlorophenoxy) propionic acid
Mecoprop	MCPP Lornox-plus	(4-chloro-2-methyl phenoxy) propionic acid
Y-Phenoxy butyric acid		
2,4-DB	Butaxone	(2,4-dichloro phenoxy) butyric acid
MCPB	Tropotox, control, Thistrol	(4-chloro-2-methyl-phenoxy) butyric acid
Misc.		
2,4-DEP	Falone	Tris [2-(2,4-dichloro-phhenoxy) ethyl] phosphate.
2,4-DES	Sesone	Sodium, 2-(2,4-dichlorophenoxy) ethyl sulphate
Nitriles		
Bromoxynil	Buctril	4-hydroxy-3, 5-dibromo benzonitrile
Dichlobenil	Casoron	2,6-dichloro benzonitrile
Ioxynil	Actril	4, hydroxyl-3, 5-di-iodobenzonitrile
Substituted Ureas		
Alkyl		
Chloroxuron	Tenoran	3-[p-(p-chlorophenoxyl) phenyl] 1,1-dimethyl urea
Diuron	Karmex	3-(3,4-dichlorophenyl) dimethyl urea.
Fenuron, TCA	Urab	1,1-dimethyl-3-phenylureamon (trichloro acetate)
Fluometuron	Cotoran, Lanex	1,1-dimethyl-3-(a,a,a-trifluorom-totyl) urea
Monuron	Telvar	3(p-chlorphenyl)-1, 1-dimethyl urea
Neburon	Bonus	3,4-dichlorophenyl butyl, methyl urea
Alkoxy		
Linuron	Lorox, Afalon	3-(3,4-dichloro phenyl)-1, methoxy-a-methyl urea.

Groups/Common name	Trade name	Structural formula
Chlorbromuron	Bromex, Maloran	3-(4-bromo-3-chlorophenyl)-1-methoxy-1-methyl urea
Unclassified		
C-288	-	
Methazole	Probe	2(3,4-dichlorophenyl)-4-methyl-1, 2,4-oxadiazolidine-3, 5-dione
Perfluidone	Destun	4-(Phenyl sulphonyl)-1,1,-trifluoromethyl sulhpono-o-toluidide

Inorganic Herbicides

Arsenic Acid

Arsenal is a non-selective total vegetation control herbicide which have both translocated and residual properties. It is mostly used in industrial sites and amenity areas where long-term weed control is needed.

Borax (Trade Name: Decahydrate): 100%

Borax is a non-selective inorganic herbicide which is used as a long-term soil sterilant in non-crop areas at a high dose (400-500 kg/ha) in anhydrous form. It is absorbed in the soil and persists for more than one year and thus renders the area unproductive. However, it leaches rapidly in high rainfall areas. In crop area, it may be used at a low rate as pre-emergence application in cotton to control *Sorghum halepense* (spot treatment), *Convolvulus arvensis*, *Cirsium arvense* etc.

Borax is mostly absorbed by roots but sufficient moisture is required to dissolve and to translocate readily from roots to apical parts of the plant. It is applied 6 kg/100 sq. m.

Borate: (Sodium Metaborate)

It is a non-selective, pre- or post-emergence herbicide which controls general vegetation on non-crop land, spot treatment of *Sorghum halepense* in cotton. The rate applied is 0.5 to 4 kg/100 sq.m.

Copper Sulphate

Copper sulphate is widely used as algaecide at 1 ppm for controlling pond slum and 2 ppm for *Chara* spp. It is applied 10 to 30 kg per cubic water flow. High dose of copper sulphate depress the photosynthetic activity. Copper sulphate is very effective to control hydrilla. Copper sulphate mixed with Fenac is more effective and economical.

Ferrous Sulphate

It induces immediate plasmolysis of cells. It controls moss on lawn, tiles, grassland and gravel at 300 kg/ha in 1 l water/m^2.

Sodium Chloride and Sodium Nitrate

The rapid action of these salts at high concentration is that of plasmolysis of the root cells and consequent interference with the water uptake by the plant.

Sodium Chlorate and Sodium Arsenite (WP 100%)

Sodium chlorate ionizes to form free radicals of heavy metal and denature plant proteins. It is applied on non-cropped lands and paths. It assimilates primarily through roots, causing necrosis. It is best to treat the soil after rain or after wetting the soil, to enhance root absorption.

Sodium arsenite acts as poison and uncouples oxidative phosporylation in respiration cycle of plants. *Najas minor* (Naidad), an aquatic algae is susceptible to sodium arsenite. *Vallisneria spiralis* (eelweed), an aquatic weed, can be controlled with sodium arsenite effectively. It is applied at 1-2 1 solution/m^2.

Sulphuric Acid

It penetrated the plant tissue rapidly and denatures the protoplasm, decomposes chlorophyll, disorganizes nuclei and plastics. It is a contact herbicide to kill dicot weeds in seeded onion. It has no soil activity.

AMS (Ammonium Sulfamate): (95%)

AMS is a non-selective contact herbicide, freely translocated in woody species on foliage application. It kills hard-to-kill broad-leaved weeds and clears of undergrowth. After application, effect it is its vapour movement to nearby susceptible weeds. It leaches in soils readily. It persists 6 to 8 weeks in tropics, but longer in dry areas. It is applied at 6% under fruit trees and at 11% (as emulsion) foliar spray on non-cropped areas.

Herbicidal Petroleum Oils

Thick oils are slow acting. They produce chronic phytotoxicty whereas high fractions of oils inflict rapid plant injury. Oils reduce the rate of photosynthesis of leaves and cause breakdown of protoplasm, thus exposing the pigment to the destructive action of light. Oils interfere with water relations of the plant. Petroleum oils impart water-soaked appearance and wilting in plans, causing eventually desiccation to death. They fill stomatal openings and occupy intercellular spaces in leaves which leads to sudden drop in photosynthetic rates.

Organic Herbicidal Compounds

Aliphatics

Aliphatics or Halo-aliphatic acids are selective translocated herbicides especially used to control perennial grasses. Dalapon and TCA belong to this group. At low rate, they cause leaf chlorosis and induce slow systemic effects; but at high rates, they cause destruction of cellular constituents and scorch the leaves intensity.

Dalapon – Sodium Dowpon: WP 85%

Dalapon is a selective translocated, pre- and post-emergence herbicide for an effective control of perennial grasses such as *Sorghum halepense*, *Cynodon dectylon*, *Agropyron repens* and also controls other grasses like *Setaria* sp., *Panicum* sp., *Digitaria* sp., *Poa annua* and emergent water weed like *Typha* sp. It contains built-in wilting agents. It is used in sugarcane, maize, sugarbeet, potatoes, jute, coffee, orchards and also on non-cropped areas. Dalapon combined with 2,4-D form a good general weed killer for both grasses and broad-leaved weeds.

Weeds are controlled effectively when applied as foliar spray on young weeds in active growth, having an adequate supply of water in soil. However, to control deep-rooted perennial grasses, repeated application at moderate rate about 5 weeks thereafter is advisable.

Its absorption and translocation is both by roots and shoots by plans. The uptake of dalapon by leaves is significant, particularly the absorption in first six hours is most important, though the absorption is continued upto 48 hours. The shoot absorption is about 10 times more rapid than TCA. Being a growth regulator, it is absorbed by roots as well as foliage and translocated readily via both apoplastic and symplastic systems freely throughout the plant. It by-passes internodes and nodes where carbohydrates are stored and the main accumulation is in growing tips and meristemetic regions. It thus inhibits elongation of primary roots within 4 hours after the treatment and the root growth is ceased within 12 hours and thus weakens regenerative capacity of grasses.

It leaches readily in soil and the persistence in the soil is very short, only 2 to 4 weeks. High rate of application causes acute plant toxicity but does not favour translocation to other plant parts resulting in quick killing of leaves take place in a localized area. At high rates in foliage, it causes rapid precipitation of cell proteins and destruction of cellular constituents, resulting in intensively scorched and brunt leaves.

TCA (Varitox) – Trichloro Acetic Acid, WP 90%

TCA is a selective translocated herbicide but at higher doses, it is non-selective and soil sterilant. It is mostly applied to the soil as pre- and post-emergence and hence its uptake by roots is most significant.

It is especially used to control perennial grasses such as Johnson grass (*S. halepense*) quack grass (*Agropyron repens*), Burmuda grass (*Cynodon dactylon*) etc. and useful for tolerant crops like sugarcane, sugarbeet, cole, crops, tomato, spinach and beans, but cereals and millets are susceptible. Hence, minimum two months interval between spraying of TCA and planting of crops is required. It is also very effective against the aquatic weed, Cattail (*Typha* sp.). Being highly soluble, TCA is readily absorbed and translocated by roots. Thereafter, it is exclusively transported via apoplastic system and accumulates in the growing tissues. It reduces both cuticle and lamina thickness in leaves, inhibits the enzyme, thus allowing the accumulation of toxic levels of free ammonia. It is readily leached from the soil because of its high water solubility.

It is applied at 5 to 10 kg/ha for selective use on crops; but for non-selective areas, the dose maybe as high as 50 to 100 kg/ha. It can be applied as its sodium salt. It is leached quickly under high rainfall and the persistence in soil is 5 to 10 weeks.

It is a strong acid. It is corrosive to iron, zinc, aluminium; also corrosive to skin and irritant to eyes and hence to be used cautiously.

Amides and Anilides

Amide herbicides comprise a diverse group soil active compounds and their biological effects vary widely. It has the important properly of inhibiting seed germination and growth of seedlings and hence majority of them are used as pre-emergence application.

It persists in soil from 7 to 8 weeks and even upto 12 weeks. While applied to the foliage, localized necrosis takes place. Amide herbicides are general growth inhibitors, particularly inhibition of root elongation. Anilides are active through soil and foliage but resist leaching in soils.

Alachlor (Lasso) EC 48%, Gr. 10%

Alachlor is a selective herbicide, mostly used as pre-emergence spray on the soil and hence sufficient soil moisture is a pre-requisite for its affectivity. In controls mostly annual grasses and certain broad-leaved weeds. The tolerant field crops include cotton, groundnut, sugarcane, vegetables, cabbage, soybean, maize and potato.

It absorbs mainly near the first node of the germinated shoots, secondarily by roots and translocated throughout the plant and then metabolized within 10 days.

It inhibits protein synthesis on susceptible plants. It inhibits growth of young seedlings by inhibiting root elongation, mitotic activity, enlargement of cortical cells and also inhibits production of hydrolytic enzymes, amylase and protease in susceptible species.

It is absorbed by soil colloids and the soil persistence is upto 10 weeks; however, no residual toxicity is found in the next crop because is breaks down within 10 to 12 weeks. The rate of application is 2 to 3 kg/ha.

Butachlor (Machete) EC 50% ai; & 5% G

Machete is one of the most recent introductions. It is used as pre-emergence selective weed killer and is mostly used to control wet-land weeds in direct seeded or transplanted rice. It is available both in liquid and granular forms; but granular form is most effective in transplanted rice to control almost all grasses and broadleaf weeds in wet land or in low standing water. Machete granule at 20 kg/ha (mixed with coarse sand) if applied 3 to 5 days after transplanting of rice, can give excellent control of variety of weeds and the field remains weed free for a long period. The weeds controlled are: *Setaria* sp. *Eleusine indica*, *Echinochloa crusgalli*, *Cyperus* sp., *Sorghum halepense*.

The liquid form is used in some tolerant crops like wheat, cotton, groundnut and *Brassica* to control *Chenopodium album* and *Amaranthus* sp.

Machete is mainly absorbed by the germinating weeds at first node stage and then translocated throughout plant parts and within 10 days, the metabolism becomes complete.

Like alachlor, it is also absorbed by soil colloids but does not persist in the soil more than 10 weeks. The applicable dose in liquid form is 1.5 to 2 kg/ha.

Propachlor (Ramrod) WP 65%

Propachlor is a pre-emergence selective herbicide, spread on soil surface or applied through furrow irrigation at preplant time. It is mostly used to control annual grasses in tolerant crops like maize, sorghum, soybean and orchards. It gives good weed control in absorptive soils. It is an inhibitor of root growth.

It is absorbed by germinating seedling, translocated throughout the plant and metabolized completely within 5 days. It inhibits protein synthesis in both root tips and shoots, just prior to cessation of growth and also decreases stearic acid content.

It is absorbed by soil colloids and persists only for 4-6 weeks. It is applied at 3 to 6 kg/ha.

Propanil (Stam F-34, Rogue) EC 34%

Propanil is a selective, contact, post-emergence herbicide, mostly used in dry-land rice to control barnyard grass (*Echinochloa crusgalli*) when this grass is very young, 2-3 leaf stage. This barnyard grass absorbs propanil at a very rapid rate and moves from leaf to the growing point and then back to other leaves. Propanil inhibits RNA and protein synthesis and blocks hill reaction in plans. As a growth inhibitor, it especially inhibits root elongation. It causes chlorosis, reduces weeds growth, destroys permeability of root and chloroplast membranes. Thus the symptoms of propanil toxicity on barnyard grass appear within 2 days after spraying. The other grass weeds affected are: *Digitaria* sp., *Cyperus rotundus*, *Eleusine indica*, and aquatic weeds like *Alternanthera* (Alligator weed). At the same time, upland rice has shown tolerance to propanil upto 2 to 3 kg ai/ha and hence the resistance of rice to propanil to inactivate it. It may be applied twice at 15 day intervals for effective control of weeds.

CDAA (Randox) Gr 20%

Randox is extensively used as selective, pre-emergence or preplant herbicide in crops like maize, beans, cotton, soybean, radish and brassicas; but used as post-emergence in onion and sugarcane. Effective weed control can be obtained by soil incorporation as pre-emergence application. It is selective against grasses like *Echinochloa crusgalli*, *Digitaria* spp. *Poa annua*, *Setaria* sp., *Avena* sp. and certain broadleaf weeds like *Amaranthus* sp., *Portulaca* sp., etc. Some grasses are particularly sensitive to Randox.

It is mainly absorbed through roots of emerged grasses and accumulates in the first internode. It inhibits root elongation and cell-division in root-tips and metabolizes rapidly. It inhibits the development of the activity of certain hormone-induced hydrobytic enzymes. It checks mitosis in root cells, stunts root growth and interferes with protein synthesis in root tips.

It is absorbed by soil colloids and persists in the soil 3 to 6 weeks. The rate of application is 3 to 5 kg actual/ha for most crops. It is irritating to skin and eyes.

Pronamide (Kerb): WP 50%

Kerb is a selective herbicide which provides most effective pre-emergence control of mostly annual grasses in orchards, groundnut, cotton, sunflower, sugarbeet etc. when the soil is kept moist. Kerb can also be used as post-emergence herbicide to control quackgrass and certain dicot species.

Kerb moves into the root zone of the weeds, translocated upward and distributed into the entire plant and is metabolized slowly. It is a strong inhibitor of mitosis. The rate of application is 1 to 2 kg/ha.

Naptalam (Alanap) Ec. 35%, Gr. 10%

Naptalam is a selective pre-emergence herbicide for the control of most of the annual grasses and broad-leaved weeds in groundnut, soybean, cucurbits and cotton. In granular form, it gives excellent weed control in established potatoes and sweet potato crops. The grassy weeds controlled are: *Setaria* sp., *Elusine indica*. Since it is applied in the soil, sufficient moisture is required after naptalam application to activate weed control properties. It inhibits root elongation and also seed germination. Seedlings which succeed in emerging are stunted and malformed. It blocks indole acetic acid (IAA) action. It leaches rapidly in light soils and hence cause no residual toxicity. It is best suited for use in heavy soils. It is applied at 2 to 5 kg/ha as liquid which in granular form at 20-30 kg/ha. Since sodium salt of naptalam is freely soluble in water, this formulation should be used.

Acrolein (Aqualin): EC 92%

Acrolein is a contact, non-selective, aquatic herbicide which is widely used to control floating weeds in the water like *Eichhornia* (water hyacinth) *Pistia* (water lettuce) etc. and also submerged weeds like *Potamogeton* (Pondweed), *Najas* etc. and *Chara* sp., *Cladophora* sp. etc. in the irrigation channels, ditches and ponds.

It is absorbed rapidly and kills the weed, by destroying the plant cells and enzyme system. The treated water is safe for irrigation and is not harmful to field crops, but toxic to fish.

The chemical is injected into water below the surface directly from its cylinder, through an extended hose at 160-450 litre/cumic for a period of 1 to 4 hours, so as to move a long distance in the flowing water and kill sub-merged weeds. It remains in the water for 2 to 3 days depending upon water temperature. In choked canals, it must be applied at several points to a concentration of 5 ppm. While handling this herbicide, gas masks have to be used. The normal rate of application is 50 ppm.

Anilines and Nitro-Phenols

Anilines and nitrophenols are used as selective herbicides. All phenol compounds give quantitative and qualitative reactions similar to colchicines and inhibit cell-division. A further manifestation of the action of substituted phenols on plant tissue is the stimulating effect on respiration.

Dinitramine (Cobex): EC 25%

Dinitramine is a selective pre-emergence or pre-planting herbicide which is sprayed in the field and incorporated after application. It is used to control annual grasses and broadleaf weeds in cotton, groundnut, soybean, beans, peas and sunflower. The grassy weeds controlled include; barnyard grass, carpetweed, crabgrass, Johnson grass, jungle rice, purslane. It prevents germination of seeds and inhibits root growth. It is metabolized rapidly by plants and degraded mostly.

It is strongly absorbed on soil and not leached through the soil. The rate of application is 0.5 to 0.75 kg ai/ha.

Nitralin (Planavin): WP 75%

Nitralin is a selective pre-emergence or pre-planting herbicide which is incorporated in the soil through mechanical means. It is useful to control annual grasses and many broad-leaf weeds from seed, particularly crab-grass, water grass and foxtails, are highly sensitive. It is used as selective herbicide in crops like rape, cotton, soybean, cole crops, cucurbits, tomatoes, groundnut and also in orchards.

It is not absorbed by plant foliage and therefore is not a post-emergence herbicide. It is absorbed only by seed or roots by inhibition of water. It inhibits cell-division. It is relatively immobile or stable in soil and leaches very slowly. The rate of application is 0.5 to 2.0 kg/ha.

Trifluralin (Treflan) EC 48%, Gr. 5%

Trifluralin is a selective, pre-emergence or pre-plant herbicide, applied on the soil and incorporated within 8 hours after application. It controls most grasses like *Echinochloa crusgalli*, *Digitaria* sp., *Setaria* sp., *Eleusine* sp., *Sorghum halepense* and certain broad-leaved weeds such as *Chenopodium album*, *Amaranthus* sp., *Tribulus terrestris*, *Portualca* sp. It is most useful in cotton and also other tolerant crops such as peas, soybean, mung, castor, cucurbits, potatoes, cole crops and in transplanted conifers. The sensitive crops are beet, sorghum, oats and spinach.

It affects seed germination and only actively growing young tissues such as root tips and coleoptiles. It is not applied to plant foliage since there is no contact activity. It induces several biochemical responses including changes in carbohydrate, lipid, nitrogen and nucleic acid content. It inhibits nodule formation and induce nitrogen deficiency symptoms. It also inhibits accumulation of cotyledonary reserves.

It is strongly absorbed on soil colloids and shows no leaching. It persists in soil for 4 to 6 months and hence it has the potential for seasonal weed control. It is applied at 0.5 to 1.5 kg/ha.

Fluochlorlalin (Basalin) EC. 48%

It is a selective pre-sowing or pre-emergence herbicide. It is very effective against many difficult annual grasses and broad leaved weeds. It is most useful in malvaceous crops like cotton, bhendi and selective to soybean, groundnut, peas, jute, onion, beans, pulses and sunflower. It keeps the crop fields weed free for 6 to 8 weeks when weeds cause maximum damage. It remains in the soil 2 to 3 months and is degraded and hence no toxic residue is left over for the next crop. The use of basalin at early stages of the crop, destroys annual weeds, helps to increase fertilizer utilization, thereby increasing the yield up to 50%.

It controls most of the weeds without harming the crops. It also does not harm leguminous crops and hence mixed crop system e.g. sugarcane inter-cropped with pulses, may be taken up.

Basalin must be thoroughly incorporated into the soil to a depth of 4-5 cm, immediately after application to avoid loss through ultraviolet light and volatilization. An irrigation must follow after sowing the crop. The susceptible grasses are; *Brachiaria*, *Cynodon*, *Digitaria*, *Echinochloa*, *Eleusine*, *Panicum*, *Phalaris*, *Setaria* and the broadleaved weeds include *Amarahtnus*, *Chenopodium*, *Euphorbia polygonum*, *Trianthema* etc. It is applied at 1 to 2 kg/ha.

Dinoseb (Pre-emergence, Dow Weed Killer): EC 48%

Dinoseb is a non-selective, contact, post-emergence herbicide which controls young seedling weeds including grasses. It can be used in orchards, maize, potatoes, groundnut, peas, lucerne, cucurbits, etc; but cruciferous crops are very sensitive. It is mostly used for post-emergence weed control in lucerne and pre-emergence in peas and beans.

It is principally a direct contact effect and no true translocation takes place. It blocks synthesis reaction and thus inhibits differentiation and growth. It is a coagulant of plant proteins. Its rapid contact action destroys the organs responsible for both apoplastic and symplastic transport. The phosphorus accumulation in leaf discs is inhibited and thus it inhibits oxidative phosphorylation in respiration mechanism of plants and thus the plant becomes energy deficient. It is a coagulant of plant proteins and hence warm weather aggravates this action of dinoseb in plants. It alters the nitrate content also.

It can leach in sandy soil. Its persistence in soil is only 2 to 4 weeks. Dinoseb sodium salt is used at 1 to 6 1 product/ha; in cereals and forage grass (1 l/ha); in orchards (2.75 l/ha).

It is dissolved in diesel oil (1 l product dissolved in 20 l diesel oil; then in 1000 l water).

PCP (Durotox)

PCP is a contact, non-selective, pre- or post-emergence herbicide which is used for the control of *Digitaria* sp., *Sorghum halepense*, *Cynodon dectylon*, *Setaria* sp., and broadleaved weeds like *Chenopodium album* and *Amaranthus* sp. It is used on tolerant crops like potato, rice, sugarcane, pineapple, citrus etc. It is also sued as a defoliant of cotton.

PCP reduces transpiration and uptake of Phosphate particularly. It also inhibits cell division and some compounds produce an over contracted late prophase.

It is applied at 2 to 15 kg/ha, using high rates on heavier soils.

Fluorodifen (Soyex, Preforan): EC 30%

Fluorodifen is used as pre-emergence herbicide broadcast to the soil surface at planting. It controls most annual grass and broad-leaved weeds. It is useful on crops like soybean, peas, beans, groundnut etc. The sensitive crops are tomatoes, subgarbeet, turnip, cucurbits, tomato, alfalfa. The rate is used at 3 to 5 kg/ha.

Nitrofen (Tok. E-25): EC. 25%

Nitrofen is a selective systemic herbicide, used both as an pre and post-emergence application. As pre-emergence, it is found very effective when the soil surface is kept moist. It is effectively used against germinating seeds and very young seedling stages and hence for its most efficient use, correct timing is extremely important. In granular form, it is commercially used as pre-emergence on transplanted rice which is most economical method of application. Its application does not harm rice root system and it has residual effect of 3 to 4 weeks. As pre-emergence, it is used in sugarbeet, groundnut, cole crops, potatoes etc. and as post-emergence in onion, carrot etc. The sensitive crops are tomato, brinjal, spinach and millets. The weeds controlled are *Chenopodium album*, *Tribulus* sp., *Portulaca oleracea* etc.

Nitrofen is a contact toxicant and hence kills susceptible seedlings when they push through the treated soil and get exposed to light. Thus the light energy is required for its herbicidal action. It causes burning of foliage and malformation of young leaves; induces stomatal closure and decreases transpiration. It inhibits respiration and photo synthesis.

It is applied at 2 to 5 kg/ha.

Arsenicals

Organo-arsenical herbicides cause chorosis, cessation of growth, browning, dehydration and death; buds fail to sprout. The arsenic acid compounds get inactivate din soil rapidly. The salts of arsenic acids are MSMA, MAA and DSMA.

DSMA (Ansar 529): EC 35%

DSMA is a selective and contact herbicide, used as pre- or post-emergence, direct spray. It is most effective on nutsedge, grasses like sandburs (*Cenchrus* sp.), Johnson grass (*Sorgham hallepense*), crab grass (*Digitaria sanguinalis*), yellow fox tail (*Setaria glauca*), *Panicum* sp., and broadleaved weeds like Cockebur (*Xanthium strumarium*) and puncture vine (*Tribulus terrestris*).

The crops like cotton, sugarcane, wheat barley, rubber, tea, coffee, cocoa, coconut, bananas and fruit trees show tolerance to this herbicide.

It is applied at 2.0 to 3.0 kg/ha.

MAA (Ansar, Weed-E-Rad): EC 35%

MAA is used extensively as a selectively post-emergence herbicide in cotton and non-cropped areas for the control of *Sorghum halepense*, *Cyperus rotundus*, *Setaria* sp., *Cenchrus* sp.

The amine salt of MAA causes accumulation of amino acids in roots of *Sorghum halepense* and thus blocks protein synthesis. Its repeated application in purple nutsedge, accelerated starch hydrolysis in the tubers and also caused very little effect on the protein content of tubers.

It is applied at 2.0 to 3.0 kg/ha

MSMA (Ansar-529, Daconate): EC 35%

MSMA is a selective, translocated, post-emergence herbicide especially used in cotton and for non-cropped areas. It controls mostly grassy weeds.

The important weeds controlled are: *Sorghum halepense*, *Cyperus rotundus*, *Setaria* sp., *Eleusine* sp., *Paspalum dilatum* etc.

The dose applied is 2 to 3 kg/ha.

Benzoics and Phenyl Acetic Acid

Chloramben (Amiben, Vegiben): EC 24%, Gr. 4%

Amiben is a soil active, selective, pre-emergence herbicide which is mostly used in soybeans to control emerged annual grasses and broadleaved weeds. The grassy weeds include *Digitaria* sp., *Setaria* sp., *Echinochloa* sp., and broadleaved weeds like *Chenopodium album*, *Amaranthus* sp. It is also used as pre-emergence herbicide on maize, cucurbits, cowpea, beans, sunflower etc. and as post-transplanting on tomato, sweet potato, pepper etc.

It inhibits root development of weed seedlings and acts on seed germination. Its absorption especially by roots, is influenced by moisture status of the germinated seed temporarily but the translocation is very much limited from roots to developing seedlings.

It moves readily in sandy soils or during rainy season. The persistence of this herbicide in the soil is 6 to 8 weeks. It is applied at 2 to 4 kg/ha.

Dicamba (Banvel): EC 50%

Dicamba is a selective, translocated, pre- or post-emergence herbicide, depending on its specific use. However, it is mostly used as a translocated post-emergence herbicide at low rates (130 g/ha) on foliar spray in conjunction with appropriate herbicides et. 2,4-D or 2,4,5-T in maize and small grains to control *Chenopodium album*, *Convolvulus arvensis*, *Sorghum halepense*, *Cyperus* sp., etc. However, the sensitive crops are soybean, beans, small-seeded legumes, ornamental sand vegetable crops. Dicamba controls difficult broadleaved weeds like Canada thistle, leafy spurge, field bindweed and brush species, and hence commonly applied at high rates (3 to 10 kg/ha) on non-cropped land. Dicamba causes molecular disorders of plasma membranes and kills the weeds.

When applied by ground spray or as granules, it is readily absorbed mainly by roots, and tends not to accumulate in the roots but is translocated throughout the plant and affects growth. It is very mobile in plants and moves in the symplast and apoplast, mainly at the growing tips. It concentrates in the young leaves at the shoot tip from where there is a general lack of movement.

Dicamba disappears fast from the soil and hence does not sterilize the soil. It is applied at 0.5-3 kg/ha. At higher rates, it is used as non-selective herbicide.

Fenac (Fenac) EC 40%

Fenac is used as a root translocated, selective pre-emergence herbicide, applied at low dose (2-3 kg/ha) in sugarcane, maize and sugarbeets to control annual and perennial broadleaved weeds at seedling stage. But its main use is made as post-emergence in non-cropped areas at high dose (8 to 15 kg/ha) against difficult broadleaved perennial weeds. It is very effective against *Striga* at 2 to 5 kg/ha. it is also used to control aquatic weeds at 4 to 20 kg/ha.

It is primarily absorbed by roots and strongly by soil colloids. It resists leaching and degradation in soils.

Carbamates, Thio-Carbamates and Carbonilates

Carbamates are particularly inhibitory to establishment of seedlings in grasses and cereals. These are compounds inhibiting cell division, causing inter-molecular precipitation and disintegration of the spindle. Certain carbamates inhibit oxidative phosphorylation, RNA and protein synthesis. Soil applied carbamate herbicides check elongation and expansion of plumules of germinating seedlings and affect the meristematic tissues. Carbamate herbicides check elongation of germinating seedlings and disrupt mitotic cell division of seedling radicles.

Thio-carbamates inhibit cell elongation and expansion of plumules of germinating seedlings. It suppresses tuber sprouts of nutsedge for a few weeks and also weakens the activity of Cynodon dactylon.

Diclormate (Sirmate): EC 80%; Gr. 5%

Dichlormate is a selective, pre-emergence herbicide used in potato, cotton, groundnut, soybean etc. The annual weeds controlled are: *Echinochloa crusgalli*, *Eleusine indica*, *Agropyron repens*, *Spergula arvensis*; and broadleaved weeds like *Chenopodium album* and *Portulaca oleracea*. It kills younger weed seedlings but does not affect the seed germination. Sufficient soil moisture is a pre-requisite for effectiveness of sirmate.

It is readily leached from the soil and hence has very little residual effect. It is applied at 1 to 2 kg ai/ha for winter crops and 3 kg/ha for rainy season crops.

Asulam (Asulox): Ec 40%

Asulam is a pre-emergence herbicide, used mostly in sugarcane to control perennial grasses like *Sorghum halepense*. It is a foliage and soil herbicide for control of Oxalis sp. (wood sorrel) and *Rumex* sp. in grassland.

It is applied as ground foliage spray and is readily absorbed by emerged foliage and roots and translocated in the meristematic regions. Thus it interferes with the process of cell division at growing points. The affect of asulam on respiration has been observed.

It is readily leached through the soil and lasts only 4 to 5 weeks. It is applied at 2 to 4 l/ha, in grasslands and Lucerne (1.6 l/ha); in sugarcane (3.5 l/ha); however, in forestry, high dose (10 l/ha) is applied.

Barban (Carbyne): EC 25%

Barban is a post-emergence herbicide applied on the ground or as the foliage spry to control mostly wild oats (*Avena fatua*) seedlings in wheat and barley. It penetrates into plant parts within hours but has very little translocation ability. In addition to wheat and barley, it is sued in sugarbeet, safflower, lentil, peas and mustard crops. It inhibits oxygen uptake and acts on auxiliary buds of wild oats. It is absorbed and held in soil upto 3 weeks. The dose used is 0.25 to 1.2 kg/ha.

Chlorpropham (CIPC): EC 40%

Chlorpropham is a highly selective pre-emergence and early post-emergence herbicide. It controls effectively many annual grassy and broadleaved weeds. The tolerant crops are alfalfa, carrot, cotton, onion, peas, sunflower etc. It is often used as pre-emergence herbicide in combination with other herbicides such as alachlor, naptalam. It can be used both in winter and summer. It is found to enhance respiration in plants. It inhibits sprouting in potatoes.

It is absorbed through the coleoptiles of emerging grass seedlings and through the roots of older plants. It is readily transported acropetally following root absorption. Its phytotoxic effects are root and epicotyl growth inhibition, suppression of transpiration and respiration. It disrupts normal cell division, strongly inhibits protein synthesis, interferes with oxidative phosphorylation and photosynthesis.

It is absorbed readily in organic matter in the soil. It is applied at 1 to 3 kg/ha, either alone or in combination with another herbicide.

Propham (IPC): WP 50%

Propham acts a selective pre-emergence, pre-planting soil incorporated herbicide and as pre-emergence on a number of annual grasses like canary grass (*Phalaris minor*), wild oats (*Avena fatua*) etc. and broadleaved weeds such as dodder (*Cuscuta* sp.), Purslane (*Portulaca oleracea*) etc.

As pre-planting application, it is sued in very tolerant crops such as lettuce, sugarbeet and forage legumes, incorporating to a depth of 8 to 10 cm for controlling grasses like *Avena fatua*. It blocks Hill reaction in plants. Post-emergence application is done only in areas where rainfall or irrigation can be assured to control selectively the susceptible weeds like *Avena fatua*. It is used only in winter months.

Propham is absorbed through underground parts, the coleoptiles of emerging grass seedlings and slowly absorbed through the leaves. It is readily transported acropetally following root absorption. It inhibits cell division. It also acts as a growth regulator.

It is applied at 1 to 4 kg/ha.

Thio-Carbamate

Benthiocarb (Saturn, Bolero): EC 50%; Gr. 10%

Benthicarb is a selective, pre-seeding or pre-emergence liquid or granular herbicide, mostly used for wetland rice to control aquatic weeds, annual grass and broadleaved weeds. The weeds controlled are: *Echinochloa crusgalli*, *Setaria* sp., *Cyperus* sp., *Eleusine indica*, *Sorghum halepense* etc. The liquid herbicide is sprayed in seeded plot in the wet-land before 3 leaf stage of paddy; it is particularly used in granular form 5 to 7 days after transplanting at 20 kg/ha. This herbicide is safe to rice. The rate of liquid is 2 to 3 kg/ha. The residual effect remains upto 30 days. The action of benthiocarb is similar to Butachlor (machete).

EPTC (Eptam): EC 75%; Gr. 10%

Eptam is a selective, pre-planting herbicide which is mechanically incorporated into the soil to a depth of 5 to 7 cm immediately after application to avoid volatile losses. The weeds controlled are: nutsedge (*Cyperus* sp.) perennial grasses such as Johnson grass (*Sorghum halepense*), barnyard grass (*Echinochloa crusgalli*), quack grass (*Agropyron repens*), in potatoes; and also broad-leaved weeds like morning glory (*Ipomea* sp.) lambs quarters (*Chenopodium* sp.) Purslane (*Portulaca oleracea*), *Solanum* sp., pigweed (*Amaranthus* sp.) etc. It is used in crops like maize, cotton, alfalfa, beans, potatoes, pineapple, citrus etc. It acts on germinating seeds and hence applied just before the seedlings are established. It is readily absorbed by roots and translocated to the leaved and epideral shoots. It inhibits growth in the meristematic region, cell-division, cell elongation and expansion of plumules of germinating seedlings. In nutsedge, it suppresses tuber sprouts for a few weeks and weaken the activity of Cynodon. It is readily metabolized by plants to carbon-di-oxide.

It is absorbed into dry soil but readily lost from the soil by volatilization when the soil surface is wet at the time of application. It is readily lost by metabolization and hence does not persist in dry sols for a longer period, 12 to 16 weeks.

The rate applied is 2 to 4 kg/ha.

Diallate (Avadex): EC 40%

Dilate is a selective, pre-plant herbicide which is mechanically incorporated in the soil at shallow depth. It provides effective pre-emergence control of wild oats (*Avena fatua*), foxtail (*Setaria* spp.) in tolerant crops like sugarbeets, maize, barley, peas, potatoes, root crops, oil crops, fruit trees etc. In wild oat, the primary absorption is through the emerging coleoptile and hence it is applied just before the weed seedlings are established. It has no action against dicots.

It is readily absorbed by roots and translocated upward to the epidermal shoots. It inhibits growth and cell division. The rate of application is 2 to 4 kg/ha.

Tri-allate (Avadex BW): EC 40%

Tri-allate is a selective, pre-planting and post-planting soil incorporated herbicide. It provides effective control of wild oats in tolerant crops like small grain crops, sugarbeets, peas, fruit crops, soybean etc. It is absorbed mainly by the coeoptile of wild oats and inhibits the cell-division an thus affects cell-elongation.

It is absorbed by soil colloids but volatilized from soil surface at high temperature, if not properly incorporated. The rate of application is 1.5 to 2.2 kg/ha.

CDEC (Vegadex): Gr. 20%

CDEC is a selective, pre-emergence herbicide. It is applied to the soil surface with a sprayer or incorporated by granular applicator. It is mainly used on the vegetable crops such as cole crops, beans, spinach, cucumber, tomatoes, bhendi etc. to control major grassy weeds like barnyard grass (*Echinochloa crusgalli*), crab grass (*Digitaria* sp.) foxtail (*Setaria* sp.) and certain broad-leaved weeds like pig weed (*Amaranthus* sp.), purslane (*Portulaca* sp.) etc. It is most suitable for weed control in sandy soils and in high rainfall areas.

It is readily taken up by roots and moves acropetally in xylem tissue. Rapid metabolism to metabolic products and carbondioxide limits accumulation in plant parts.

Moderate rainfall after application of CDEC to the soil surface enhances it performance. However, its activity is decreased by volatility losses at high temperature. It persists in the soil generally for 3 to 6 weeks. The rate usually applied is at 2 to 4 kg/ha.

Molinate (Ordram-72): EC 72%; Gr. 75%

Molinate is a selective pre-plant, soil incorporated herbicide, mainly used against *Panicum* sp. and to a lesser extent against *Typha* sp. or and also for the control of barnyard grass (*Echinochloa* sp.) and other weeds like nutgrass (*Cyperus*). It is applied on the dry soil and incorporated with disc harrow before the field is flooded for planting of water-seeded rice or after the watergrass is just emerged in water-seeded rice, as post-emergence.

It is readily absorbed by the emerging seedling shoots and then translocated upwards. It is readily metabolized by plants to carbon dioxide to a large extent. Its phytotoxic effect is abnormal emergence of leaves for the cleoptiles.

It is absorbed on to dry soil, but readily volatilized from wet soil. It breaks down in the soil very slowly but does not persist in the soil. The rate of application is 2 to 3 kg/ha.

Bentazon (Basagran): WP 48%

Basagran is a pre-emergence herbicides, mainly used in cereals from 2 to 3 leaf stage, to control dicot seedling weeds. It is also used in beans, peas, potatoes and soybeans.

SWEP

Swep is a pre-emergence herbicide which is most effective for selective control of germinating weeds in rice and certain legumes.

It is applied at 4 to 6 kg/ha.

Glyphosate (Round up): EC 41%

Glyphosate is relatively non-selective new compounds, translocated with great efficiency into the rootstocks and rhizomes of perennial plants. It is very effective on deep-rooted perennial grasses like *Agropyron rpens* (quack grass), sedge and broad-leaf weeds destruction of semi-submerged and floating aquatic weeds. It is applied only as post-emergence spray to the foliage of vegetation to be controlled. It is an extremely effective total herbicide and hence drift on the cultivated crop should be carefully avoided. For its successful use: (i) at least 6 hours freedom from rain after application (ii) adequate foliage to receive spray chemicals and (iii) target weeds should be at active growing stage for efficient translocation, are the pre-requisite.

It is assimilated only through active portion of the plant within 3 to 4 hours, and translocated throughout the aerial and underground portions of the perennial weeds, resulting in failure of regrowth and complete destruction.

It is firmly absorbed in all types of soil and the leaching is very slow. It is relatively non-persistent in the soil and thus crops can be planted or seeded directly without any harm.

The rate used is 1 to 2 l/ha for perennials and 5 to 10 l product/ha for aquatic weeds. For annual weeds, 1 l/ha; for biennials 2.2 l/ha and from perennials 4.3 l/ha.

Heterocyclic Compounds

Bipyridylium

Bipyridylium herbicides are active only when applied to leaves. They also inhibit net photosynthesis when applied to the roots. Both diquat and paraquat bring about rapid desiccation of treated vegetation and discoloration withinan hour of their application, followed by wilting within 20 hrs in sunny condition. Both diquat and paraquat bring about a rapid desiccation of treated areas.

Diquat (Reglone): EC 20%

Diquat is a non-selective, post-emergence and contact herbicide. It is used only as a non-crop weed killer and mostly used on aquatic weed control. It is also used as pre-harvest desiccant of seed crops and trash control in sugarcane. On sub-merged aquatic weeds, it is applied below the water surface or by pouring it into the water while moving slowly in a boat. In case of floating aquatic weeds, it is applied by wetting the foliage thoroughly. Diquat can be mixed with 2,4-D dalapon, simazine, bromacil, diuron etc. it can be used as directed spray around tea, coffee, fruit and forest trees.

It is very rapidly absorbed by foliage and also very resistant to removal by water. It causes rapid desiccation when applied as preharvest desiccant, followed by necrosis and ultimate death of leaf. Its action is much faster in the light than in the dark and hence the chemical is not activated until brought into the light. It interferes with oxidation-reduction reaction in the plant. It is translocated less readily in grasses. Its toxicity is increased with post-treatment light illumination.

An important property of diquat is its very rapid and complex inactivation by soil. It has also very long persistence in the soil. It is applied at 0.5 to 1.5 kg/ha.

Paraquat (Gramoxone): EC 20%

Paraquat is a contact non-selective, post-emergence weed killer containing built in wetting agents. It is herbicide for non-crop areas, road side, rail tracts and pasture renovation. It is also used as crop desiccant or defoliant on cotton, potatoes, soybean pineapple, banana, grapes, cassava. It is applied against aquatic weeds also.

It is applied only to foliage and the contact action is very rapid which is often visible within a few hours and is followed within 24 hours by wilting of the leaves and green tissues, if comes in contact with. Most of its activity is by local absorption and local systemic action. Plants are killed rapidly and its rapid action occurs in bright sunlight and the toxicity is increased in light. It can be used as directed spray around fruit trees, tea, coffee, since it is not absorbed in woody portion of these species.

It is completely inactivated in the soil but has very long persistence and hence cannot be moved by washing. It controls weeds at early stage of their growth, thereby reducing competition of water and nutrients. It is applied in very low dose 0.25 to 0.75 kg/ha for desiccation medium dose 0.5 kg/ha in pasture renovation and for general weed control, it is applied at 1 kg/ha.

Pyridines

Picloram (TORDON): EC 25%

Picloram is a selective, translocated, pre- and post-emergence herbicide, mostly used in non-crop areas for the control of woody perennials. It is the most powerful growth regulator. Since most grasses and cruciferous plants are resistant to picloram, it acts mainly against broadleaved weeds viz. field bind weed (*Convolvulus* sp.) Canada thistle (*Cirsium arvense*) etc. it is applied in liquid and granular form, granular application is most effective in most soils. Picloram is an auxin type herbicide which develops phytotoxic symptoms and stimulates secretion of salts into root xylem. Being an extremely mobile compound, it is readily absorbed by the foliage and roots and thus translocated in both the phloem and xylem vessels and finally accumulated in the active region. It produces abnormalities throughout the root system, causing severe injury to root-primodia, swelling and destruction of parenchyma tissues. It is very effective, particularly against woody species of broadleaved weed because of its ready uptake by leaves and roots and easy translocation to the top of the plants where it concentrates on shoot-tips and young leaves. Spraying of picloram on the vegetation causes curling of stem tips, cracking of plant parts and death of growing points. The inherent high toxicity of picloram is the main reason for its good kill of deep rooted perennials.

It is very stable in soils but disappears rapidly under warm and humid condition. It has long lasting residual activity (100 g/ha).

It is applied at 1.5 to 8 kg/ha as granules in the soils and at 1 to 3 kg/ha as liquid spray against difficult perennial weeds like Canada thistle and bind weed.

Pyridazines

Maleic Hydrazide (MH): EC 30%

MH is used as growth regulator and also acts primarily as a growth inhibitor against grasses. Its primary uses include sprout inhibition in stored onion and potatoes; sucker control on tob*acco and herbicidal activity on quack grass (Agropyran repens*), wild onion (*Allium canadense*) and wild garlic (*Allium vineale*).

It is used as foliar spray over the top. MH causes phloem necrosis and disturb normal translocation of food. At high rates, MH leads the plants to death. It is slowly absorbed within 24 hr. by quackgrass but most effective absorption occurs when soil moisture is at field capacity and at high relative humidity. Once absorbed, it is freely translocated to active growing points in the plant. It becomes fixed within the plant and is not metabolized. It persists for 2 to 5 weeks in soils.

It is applied at 2.5 to 5 kg ai/ha.

Uracils

Bromacil (Hyvar-X): WP 80%

Bromacil is mainly a soil sterilant and mostly used on non-cropped areas for the control of mainly grasses like Johnson grass (*Sorghum halepense*), Bermuda grass (*Cynodon dactylon)*, foxtail (*Setaria* sp.), wild oats (*Avena fatus*), crabgrass (*Digitaria* sp.) and knot grass (*Paspalum* sp.). It is combined with diuron to control weeds in non-cropped areas most effectively.

Bromacil is also used as selective soil-applied herbicide at low rate (2 to 4 kg/ha actual) as pre- or post-emergence in pineapple and citrus orchards, to control seedling of monocot weeds. It has to be sprayed on soil surface at active growing period of weeds.

It is a strong inhibitor of Hill reaction and causes necrosis of root meristems. It is primarily absorbed through roots and specifically inhibits photosynthesis. Its activity is found approximately 83% in the roots and 17% in the stems and leaves of the weed plant. It is destroyed by soil microbes.

It is slowly absorbed on soil colloids but persists more than one season, when high sterilant rates are applied.

It is applied at 2 to 3 kg/ha for pineapple, 3 to 6 kg/ha in citrus orchards; 5 to 10 kg/ha to control woody perennials and 12 to 20 kg/ha as soil sterilant to control hard-to-kill perennials like *Sorghum halepense*, *Cyperus rotundus*, *Cynodon dactylon*. It persists in the soil 20 to 24 weeks at low rates and at high rates, more than a year. It is applied both in granular and liquid form.

Terbacil (Sinbar): WP 80%

Terbacil is a non-selective herbicide, widely applied to non-cropped areas; but also sued for selectively control of annual weeds in sugarcane and orchards. The grassy weeds controlled are barnyard grass (*Echinochloa* sp.) blue grass (*Poa annua*) and Johnson grass (*Sorghum halepense*). It is also used on roadways.

It is sprayed on soil surface, just before active growth of weeds as pre-emergence on sugarcane. It is rapidly absorbed through root system and is translocated upward into the leaves.

It has slow absorption in soil colloids and persists in the soil 5 to 6 months. It is usually applied at 1 to 3 kg/ha and in non-cropped areas, at 3 to 8 kg/ha to control yellow nutsedge (*Cyperus* sp.), Bermuda grass (Cynodon) etc.

Bentazon (Basagran): WP 48%

Bentazon is a selective, post-emergence and contact herbicide. It is selectively controls many broad-leaved weeds and sedge by contact action in most cereals and leguminous crops. The broad-leaved weeds controlled are: *Amaranthus* sp., *Convolvulous* sp. etc.

It is very effective after foliar absorption. Acropetal movement occurs rapidly through xylem tissue after root-absorption.

It is applied at 0.5 to 1.5 kg/ha.

Triazines

Triazines are inhibitors of growth of most plant organs, leaves become chlorotic and finally necrotic symptoms develop. This group has an adverse effect on carbohydrate content of plants, reduce the amount of glucose, fructose and sucrose and thereby inhibit photosynthesis. Of this family, atrazine and simazine are best known which accumulate in chloroplasts and also hamper germination of susceptible plans. They increase total free fatty acid content. The S-triazine herbicides are readily absorbed and translocated throughout the plants. The order of toxicity of this group is atrazine > simazine > propazine > atratone > prometone > prometryne. There are three types of S-trizine herbicides namely; chloro-s-triazines (Atrazine, propazine, simazine), methoxy s-triazines (prometone etc.) and methyl-s-triazines (ametryne, prometryne, terbutryn). Chloro-s-triazines are translocated by roots; while methoxy and methyl-s-triazines are active through both roots and shoots. However, methoxy-s-triazines are mostly non-selective, long persistent and hence most suitable against deep rooted noxious weeds. Methyl-s-triazines accumulate in chloroplasts and block light, hamper germination and inhibit photosynthesis. The mature leaves become chlorotic first with margin inward, and the youngest leaf last.

Atrazine (Atrataf, Residox): WP 50%

Atrazine is widely used as a pre-emergence of early post-emergence selective herbicide; but pre-emergence use is generally preferred. It is active both in roots and shoots; on shoots, it has contact action against tender weeds. It has become the standard herbicide in maize at low rates and hence commercially used; because of its outstanding selectivity to maize, giving excellent control of established weeds. It can be used in certain crops like sugarcane, sorghum, pineapple. In cropped areas, the main dicotyledonous weeds controlled are; *Chenopodium album*, *Portulaca* sp; and grassy weeds like *Echinochloa* sp., *Cyperus* sp., *Setaria* sp., *Avena fatua*, *Agropyron repens* etc.

It is also extensively used as a non-selective herbicide for vegetation control on non-cropped areas at higher rates.

It is mostly absorbed through roots as pre-emergence application in maize and enters the plant parts rapidly, then accumulates in the apical meristem and in chloroplasts of susceptible weed species. It is more effective than simaine in low soil moisture because of its better water solubility. It inhibits photosynthesis and enzyme reaction. It causes foliar chlorosis showing margin inward, and thus mature leaves show the symptoms first. It ceases cambial activity and reduces thickness of cell wall. It prevents the opening of stomata in presence of the light. The inhibition of carbohydrate synthesis weakens the plant and the root system is also weakens, resulting in subsequent death. It also hampers germination of susceptible weeds for which it requires moisture to move down to the root zone of germinating seeds. It is highly soluble in water and hence need less moisture for effective weed control and thus preferred drier conditions in which maize is grown.

It is rapidly absorbed in the soil and remains active for over one year under dry condition and hence it is most useful in low rainfall areas.

It is applied at 0.5 to 2 kg ai/ha. High rates 2 to 4 kg/ha are used in non-cropped areas.

Simazine (Gesatop, Prinlep): WP 50%, Gr. 20%

Simazine is a pre-emergence, selective and systemic soil herbicide, mostly used for control of broadleaved weeds and certain grassy weeds in crops like maize, sugarcane, pineapple and citrus orchards. However, it is susceptible to certain crops like sugarbeets, tobacco and vegetables. The weeds controlled are: *Chenopodium album*, *Amaranthus* sp., *Digitaria* sp., *Setaria* sp., *Agroprogon* sp., *Poa anua* etc.

It is mostly used to control weeds in deep-rooted crops like citrus, fruit crops, coffee etc. It can be used as soil sterilant for treatment of ditch banks and drainage channels and as algaecide for fish ponds and hatcheries.

Applications are made on soil in shallow incorporation prior to weed emergence and also prior to planting of certain crops. It is absorbed mostly through roots and translocated to other parts of the plant through xylem and then accumulates in the apical meristems. It has no foliar penetration.

It is readily absorbed through roots on clayer soils, having very low water solubility and downward leaching movement. It persists longer in the soil under dry condition, however many rotational crops can be planted one year after application. At relatively higher rates, it acts as effective soil sterilant. It is applied at 1 to 3 kg/ha for selective weed control. As algaecide, the effective dose is 1 ppmw. In industrial areas and non-cropped land, higher dose of 10 to 20 kg/ha can be applied as soil sterilant. It leaches readily in sandy and gravelly soils. If it persists longer in heavy readily in sandy and gravelly soils. If it persists longer in heavy soils, the rotation crop of wheat or other crops can injure badly.

Propazine (Gesamil, Milogard): WP 50%

Propazine is a pre-emergence selective herbicide, widely used to control both broadleaved and grassy weeds, especially in sorghum (jowar). It is sprayed prior to emergence of weeds or sorghum. The combination of propazine plus terbutryne is used for weed control in sorghum. The sensitive crops are sugarbeet and vegetables. It is applied on soil as shallow incorporation. It has very like post-emergence activity.

It is absorbed mainly through roots and is translocated acropetally in the xylem. It accumulates in the apical meristems and younger leaves, causing death of plants. It inhibits photosynthesis.

It is more readily absorbed on clay soils and has long residual effect upto 12 months, particularly under dry condition.

It is applied at 0.5 to 2 kg/ha in field crops and upto 5 to 7 kg/ha in vine yards.

Ametryne (Gesapex): WP 80%

Ametryne is a selective pre or early post-emergence herbicide which control mostly grass such as *Digitaria* sp. (crabgrass), Setaria sp. (foaxtail); and certain broadleaf weeds like *Portulaca* sp., *Chenopodium* sp., *Amaranthus* sp. The tolerance crops are pineapple, sugarcane, bananas etc. It is also used as post-emergence spray in maize and as a desiccant of potato vines because of its considerable contact activity. Its shallow incorporation in the soil as pre-emergence proves best.

It is absorbed through both roots and foliage, but translocated only acropetally in xylem. It penetrates the foliage rapidly. It also absorbs through roots and translocated acropetally through xylem and accumulates in the apical meristems. It inhibits photosynthesis. It is readily absorbed on clay soils and its residual effect remains several months under tropical conditions. The dose applied at 2 to 3 kg/ha for short duration crops; 3 to 4 kg/ha for sugarcane, pineapple etc. and 2.5 kg/ha for bananas.

Prometone (Pramitol): EC 25%

Prometone is a non-selective pre or post-emergence herbicide which controls both broadleaved and grassy weeds on non-cropped areas. It is active through both root and shoot. It is applied on the soil either before or after weed emergence. It is absorbed through the roots and hence adequate moisture is required to move this into the root zone. The higher rates are used to control Johnson grass, Bermuda grass, bindweed and wild carrot.

It is translocated from roots to stems acropetally. It is more readily absorbed on clayey soils. It is applied at 6 to 20 kg/ha.

Promatryne (Gesagerd): WP 50%

Promatryne is a pre-emergence selective systemic herbicide which is mostly used on cotton and also on potatoes and soybeans. It is very effective on annual broadleaved weeds such as *Amaranthus* sp., *Solanum nigrum*, Convolvulus sp., Portulaca sp., and also some grasses like *Echinochloa*, *Digitaria* sp. At high rates, it may be used in non-cropped areas.

It has contact action. It also activates through root absorption, but penetrates the foliage rapidly. It translocated from roots through xylem and accumulates in the meristematic regions.

It is more rapidly absorbed in clayey soils but does not have residual effect. It is applied at 1.5 to 3 kg/ha.

Terbutryn (Igran): WP 80%

Terbutryn is a selective, per- or post-emergence herbicide for control of annual grassy weeds and certain difficult broadleaf weeds. It is mostly used as pre-emergence in sorghum and wheat. It is also selective to pea and potato.

It is absorbed through both foliage and roots. It is translocated from roots through xylem and accumulates in the apical meristems.

It is readily absorbed on clay soils containing organic matter. Its residual activity is from 3 to 10 weeks depending upon rate used, soil type and weather. It is applied at 1.0 to 2.0 kg/ha.

Metribuzin (Sencor): WP 70%

Metribuzin is used as pre-emergence or early post-emergence herbicide for the control of many broadleaved weeds and also grasses in soybean, potatoes, tomatoes, sugarcane and fruit crops. The sensitive crops are sweet potato, cole crops, onion, cruciferous crops, peas etc. The major deep-rooted weeds controlled are: *Xanthium* sp., *Cassia tora* etc.

It is absorbed through leaves and roots in post-emergence

application. While begin absorbed through roots, it is translocated upwards and acts as a systemic herbicide. It inhibits photosynthesis and has long acting soil action.

It is readily absorbed in clay soil and activates at very low soil moisture conditions and persists in the soil for 10 to 12 weeks. It is applied at 0.25 to 1.0 kg/ha for soybean, tomato and potato and 2.0 to 6.0 kg/ha for sugarcane.

Misc. Heterocyclic Compounds

Aminotriazole

Amitrole (Weedazol; Amizol): WP 50%

Amitrole is a non-selective, translocated pre or post-emergence herbicide, mostly used in non-cropped areas, against mainly perennial broadleaved weeds like *Amaranthus viridis*, *Euphorbai* sp., *Cirsium arvense* (Canada thistle), *Rhus radicans* (poison ivy), *Calotrpis procera* (milkweed); certain annual grasses/sedges *Agropyron repens*, *Cynodon dactylon*, *Cyperus rotundus*, *Poa annua*, *Allium* sp., *Setaria glauca* and also some aquatic weeds *Echhornia crassipes*, *Typha* sp. etc. Most of them are sensitive, if contacted by spray.

It is used as pre-plant treatment in cotton and orchards and also used as a defoliant in cotton.

It is readily absorbed by the foliage. It is highly mobile and hence readily translocated to rhizomes of perennial weeds, deep into the soil and injure auxiliary buds. Through foliar application, it is readily translocated to the meristematic tissues through symplastic and apoplastic systems. It inhibits chlorophyll formation, affects formation of chloroplasts in younger leaves, causing photosynthetic disruption and thereby inhibits regrowth. It produces phytotoxicity symptoms, showing bleaching first in young and then older leaves.

It is readily absorbed on organic soils but has very short residual life (2 to 4 weeks only). The rate of application is 2 to 10 kg ai/ha. A combination of amitrole with 2,4-D or 2,4,5-T is compatible.

Endothall (Endothall): EC 55%

Endothall is a pre- or post-emergence herbicide used as aquatic herbicide (safe to fish) and also as a defoliant. It is selectively toxic to plants. It is both root translocated and shoot contact herbicide and hence precaution has to be taken to prevent drifts on cultivated crops.

It is selective to certain crop plants like sugarbeet, onion, potatoes, soybean, groundnut and cole crops. It can also be used as general weed killer while mixed with AMS, a non-selective foliage active herbicide. Endothall persists in soil for 8 to 10 weeks. It is used for the control of submerged weeds and pond scums in confined water at 1 to 5 ppmw. Treated water is safe to use after 7 days. It irritates skin and eyes.

Oxadiazon (Ronstar): EC 25%

Oxadiazon is a pre- or post-emergence herbicide which is mostly used on rice and also in soybeans, onion, potatoes and orchards. It kills mostly broadleaved weed and a few annual grasses.

It is most actively absorbed by foliage but strongly absorbed by soil colloids when used in rice. It persists in the soil 10 to 12 weeks and thus keep the dryland rice crop free from weeds. It is applied at 0.75 to 4 kg ai/ha depending on weed population.

Pyrazon (Pyramin): WP 45%

Pyrazno is a pre-emergence or early post-emergence herbicide which controls mostly annual grasses in sugarbeet. It is absorbed through roots and subsequently translocated to all plant parts through xylem vessels.

It is applied at 2 to 4 kg/ha but in sandy loam soil, the lower rate has to be applied.

Hormone-Type herbicides

Hormone type phenoxy-acids have been extensively used as herbicides. Many formulations containing salts, amines and esters of the phenoxy, alkanoic acids have been introduced because the derivatives are taken up by plants more readily than free acids. They act as uncouplers and inhibit oxidative phosporylation. The harmful effect of phenoxy herbicides on wheat, barley and oat during very early phase and before pre-heading phase are very spectacular.

a. Phenony Acetic Acid: WP 80%

2,4-D: (Weedone – ethyl ester: Fernoxone, Weedox, Tafacide – Sodium salt; Dedweed-40, amine salt; Decamine – special (formulation)

2,4-D is a systemic, selective, post-emergence translocated herbicide which is widely used to control broadleaved weeds in cereal crops, sugarcane and also in non-cropped areas. The broadleaved weeds controlled are: *Chenopodium*, *Convolvulus*, *Amaranthus* sp. etc.

The common forms of 2,4-D available are: Sodium salt, amine salt and ester formulations. The sodium salts are water soluble powders; amine salts are water soluble liquids; but ester-compounds are mostly soluble in petroleum oils and no easily washed from spread leaves and hence is more toxic to plants and thus used for controlling woody perennial weeds. The ester formulation of 2,4-D (Weedone) produces more chromosome aberrations. 2,4-D first acts as hormone and stimulates growth of epidermal and cortical cells; at the second stage, the cambium cells are stimulated; respiration is accelerated, carbohydrates are depleted and morphological mal-formation occurs; at the third stage, toxic action is maximum, photosynthetic activity ceases completely and growth is arrested, transport of minerals is impacted; roots are emerged from the swelling points and the plants are about to die. At the final stage, plans are killed. 2,4-D thus kills the plant in four stages: stimulation, proliferation, necrosis and death.

Fenoxone is a translocated selective herbicide for the control of broadleaved weeds and sedges, containing built-in wetting agent.

2,4-D is mostly effective during seed germination as it is absorbed from the soil by young seedlings with fatal results. Once it has contacted the plant, it is readily absorbed and translocated within the phloem through the tissues and causes disruption of its phloem tissues and consequent dislocation of photosyntheates symptoms and kills even the perennial weeds efficiently. It accumulates in the meristematic regions principally and normal cell division ceases, affects respiration rate, increases turgor pressure of cells, causes abnormal growth and close of stomata, resulting in death of plants.

It is absorbed by leaves and moves in the apoplast and symplast. In most species, its downward movement is slight and accumulates mostly in apical tissues.

It is quickly leached in sandy soils and it remains longer only dry soil. It is applied at 0.5 to 2 kg ai/ha but for non-cropped areas, the dose may be increased upto 4 kg/ha. In cereals the dose would be 0.8 amine, 1.0 kg sodium and 0.5 kg ester salts while in floating and semi-submerged dicot aquatic weeds, 2 to 3 kg ai/ha.

2,4,5-T (Brush-killer): EC 48%

2,4,5-T is used as post-emergence, translocated herbicide, generally used in non-cropped areas, mainly for the control of woody perennial weeds and also applied to individual stumps. The major woody perennial weeds controlled are: Poison ivy (*Rhus radians*), Wild oat (*Quercus* sp.) and black berry (*Rubus* sp.).

It is absorbed through roots and foliage and then translocated to plant and moves to the phloem freely. There is no build up in soil. High humidity decreases upward movement of herbicides. When soil moisture is low, the translocation of 2,4,5-T is reduced because of lack of carbohydrate supply for phloem.

It is applied only at 1 to 3 kg/ha normally; but for woody perennials, it is applied at 6 to 16 kg/ha as foliar spray in repeated application. Brush killer containing a mixture of 2,4-D and 2,4,5-T is also sued which is most effective.

MCPA (Chiptox): WP 75%

MCPA is a selective, post-emergence, translocated herbicide which is mainly sprayed on foliage of broadleaf weeds. It is mostly used on cereals (barley, wheat, sorghum, rice, maize, and oat) and also in potatoes and beans. The weeds controlled are: *Tribulus*, *Xanthium strumarium*, *Portulaca*, *Ambrosia*, *Convolvulus* sp., etc. It is used for aquatic weed control.

It is absorbed readily through roots and readily translocated in plant parts. The absorption is increased by wetting agent.

It is readily leached from soil but persists upto 6 months in dry soil. It is applied at 0.25 to 2 kg/ha.

b. ∝-Phenoxy Propionic Acid

Dichlorprop (2,4-DP): EC 50%

Dichlorprp is a post-emergence translocated herbicide, mainly used for the control of woody perennials and herbaceous dicot weeds as individual stumps treatment. It is used for brush control in non-agricultural land. It is applied at 9 kg/ha in dormant bush.

Fenoprop or Silvex (Weedone, 2,4,5-TP)

Fenoprop is a selective, post-emergence translocated herbicide, which is used to control of woody broadleaved weeds; difficult surface and emersed aquatic weeds particularly waterlity, alligator weed and spike rush in ditches and riverbanks. It is more effective than 2,4,5-T. It is also used to control certain broadleaved weeds in sugarcane.

It is absorbed through roots and foliage and then readily translocated along with food materials in the phloem. It disturbs the normal process of cell differentiation, ester-forms are more effective as foliage spray than salts and the ester formulations have low solubility in water. Both silvex and 2,4,5-T are more persistent than 2,4-D.

It is applied 1 kg/ha in sugarcane, 5 to 10 kg/ha as bush spray and 4 kg/ha in water against aquatic weeds. It can be injected into the water at 2-4 ppm w in stagnant water because it needs about 2-3 weeks for the contact period to kill the weeds. It should not be used in moving water.

Mecoprop (MCPP): EC 50%

Mecoprop is a selective, post-emergence translocated herbicide which is used as broadleaf weed killer, particularly for control of knotweed (*Polygonum* spp.), pigweed (*Amaranthus* sp.) lambs quarters (*Chenopodium album*) etc. in cereal crops. It is most useful for control of weeds in lawn and turf.

Like MCPA, translocation of mecoprop takes place from the foliage into roots. Drought produces a delay in weed control action. It is applied at 1.5 to 2.5 kg/ha in cereals and at 2.5 to 3.5 kg/ha in lawn and turf.

γ-Phenoxy Butryric Acid

2,4-DB (Butaxone): EC 24%

2,4-DB is a selective, translocated post-emergence herbicide which controls certain broadleaved weeds in lucerne, soybean and forage legumes. The main weeds controlled are: cocklebur, *Chenopodium album*, *Xanthium* sp., *Convolvulus* sp., *Solanum* sp., *Amaranthus* sp. etc.

It is absorbed through foliage, readily translocated in plant system and then oxidized. Photosynthesis is inhibited and the phloem ceased to function.

It applied at 25 l/ha.

MCPB (Tropotox, Cantrol, Thistrol): EC 40%

MCPB is a selective, translocated post-emergence herbicide which is a broadleaved weed killer. It is more selective than MCPA and

2,4-D. It is more effective to *Cersium arvense* (Canada thistle). Other broadleaved weeds controlled are: *Amaranthus* sp., *Ipomoea* sp., *Chenpodium* sp., *Portulaca* sp. It has high selectivity, since it affects those plants which metabolize its phenoxy-butyric acid to phenoxy acetic acid one. It is applied at late tillering phase of cereals.

It is absorbed through leaves and readily translocated throughout the plant parts. It persists on the dry soil from one to 6 months.

It is applied at 0.5 to 1.5 kg/ha depending upon the stage of the weed.

2,4-DEP (Falone): WP 25%

2,4-DEP is a selective per-emergence herbicide which is used to control annual grasses and broadleaved weeds in crops like potatoes, groundnut, etc. The sensitive crops are cotton, tobacco, tomato, onion and grapes. The weeds controlled are: *Amaranthus* sp., *Chenopodium* sp., *Ambrosia* sp., *Digitaria* sp., *Echinochloa* sp., *Setaria* sp., etc.

2,4-DEP required sufficient moisture for activation. It is strongly absorbed by root tissues, readily translocated by the plant and accumulates at the meristematic regions.

It persists in the soil 3 to 8 weeks. It is applied at 1 to 6 kg/ha.

2,4-DES (Sesone): WP 98%

Sesone is a selective, pre-emergence herbicide which controls both broadleaved and grassy weeds on groundnut, maize, potatoes and orchards. However, it is mostly effective on groundnut. The weeds controlled are: *Chenopodium* sp., *Convolvulus* sp., *Amaranthus* sp., *Portulaca* sp., and grasses like *Echinochloa* sp., *Digiteria* sp., *Setaria* sp.

It moves readily through the tissues and accumulation occurs at the apical regions. It becomes an effective herbicide while in contact with the soil. It requires sufficient moisture for activation and to control of germinating seeds. It is applied at 2 to 5 kg/ha.

Nitriles

Bromoxynil and Ioxynil cause necrotic patches in leaves and slowly these patches joint to wither away the whole leaf within 1 or 2 days. They are inhibitor of Hill reaction.

Bromoxynil (Buctril): EC 35%

Bromoxynil is a pre-emergence content herbicide which is applied to the soil at post-planting of the crop. It is used on small grains and linseed to control certain broadleaved weeds. It is also applied in non-cropped areas, industrial sites, rail-road sides.

The oil soluble amine and ester of Bromoxynil resist removal from foliage and have no movement once absorbed. It is a inhibitor of photosynthesis and respiratory activities. It is applied at 0.2 to 0.5 kg/ha.

Dichlobenil (Casoron): WP 45%

Dichlobenil is strictly a pre-emergence herbicide applied for the soil at post-planting to crop and pre-emergence to weeds. It is used for aquatic weed control; it also inhibits sprouting of potatoes. It is strong inhibitor of germination of weed seeds, rhizomes and tubers; it should be applied to weed-free soil for efficiency.

It is used on crops like alfalfa, citrus, fruit orchards etc. The weeds controlled are: *Amaranthus* sp., *Chenopodium album*, and water weeds such as *Potamogeton* sp., *Najas* sp., *Eichhornia crassipes* etc.

Some of the sprouts of perennial weeds like *Cyperus rotundus*, *Cynodon dactyton*, *Agropyron repens* etc. is stopped but root systems survive which eventually reproduce new shoots. Dichlobenil inhibits synthesis of auxins and impedes amylase activity in germinating seeds. It causes destruction of phloem, cambium and parenchyma tissues; acts primarily on growing points and rapidly by roots. Main action is the inhibition of germination. Dichlobenil treated seedlings fail to emerge. It is absorbed for the soil by root system and translocated very fast into the aerial parts.

It is fairly persistent in soil and remains effective for about 2 to 6 months. The rate used at 1 to 3 kg in field crops, 2.5 to 5 kg/ha for orchards; and 5 to 10 kg for aquatic weeds and 10 to 20 kg/ha for total weed control on non-cropped areas.

Ioxynil (Actril): EC 25%

Ioxynil is a pre-emergence herbicide applied to the soil. It controls broadleaf weed seedlings in cereal grains including wheat and barley. It is an inhibitor of photosynthesis and respiratory organs. It needs light to take effect. It is applied at 0.5 to 1 kg/ha.

Substituted Ureas

Substituted ureas are mostly soil active herbicides, selective at low rates and as soil sterilant at high rates. Pre-emergence control of annual and perennial grasses is possible, because the substituted ureas are taken up both by foliage and through the roots. Their persistence in the soil depends on water solubility.

Urea herbicides inhibit photosynthetic activity and reduce sugar content in treated plants and also affect adversely on carbohydrate levels in plants. The most common substituted ureas are diuron and monuron which inhibit phosphorylation in plants. The substituted urea compounds interfere nitrogen metabolism. They are rapidly absorbed by roots and accumulate in tops within 48 hours but having no translocation. Thus, both root and shoot growth are severely affected.

Both monuron and diuron are used to block in hill reaction and inhibit non-cyclic phosphorilation.

Chloroxuron (Tenoran): WP 50%

Chloroxuron is a selective, pre-emergence herbicide which controls most annual grasses and broadleaved weeds in a number of vegetables (carrot, pea, potato, bhendi, beans, tomatoes, chili, onion, garlic) soybean, groundnut, citrus orchards, under sufficient soil moisture condition. It also controls moss on lawns. Pre-emergence is done at planting and post-emergence just after weed emergence. The grass weeds controlled are: *Setaria* sp., *Eleusine* sp., *Digitaria* sp., *Echinochloa* sp.; and broadleaf weeds are *Amaranthus* sp., *Chenopodium album*, *Xanthium* sp., etc. It has considerable foliar activity and hence controls young weeds effectively. It also controls certain aquatic weeds like *Potamogeton* sp., *Chara* sp., *Najaur* sp. The sensitive crops are sugarbeet and cole crops.

It is absorbed mainly through apical region of roots along with water movement. It is also absorbed strongly on soil particles and does not normally leach. It is applied at 2 to 3 kg/ha for soybeans and up to 6 kg/ha for onion and carrot.

Diuron (Karmex): WP 80%

Diuron is relatively non-selective herbicide and is mostly applied as soil sterilant. However, at low rates, it acts as pre-emergence herbicide and selectively controls germinating broadleaved and grassy weeds in crops like tea, coffee, bananas, coconut, sugarcane, cotton, pineapple, grapes etc. It has been used as pre-emergence in cotton for long. At high rates, it acts as non-selective general weed killer and also as soil sterilant in non-crop areas. It is an effective weedicide for the control of many annual and perennial grasses and herbaceous weeds. The main weeds controlled are: *Digitaria* sp., *Echinochloa crusgalli*, *Sorghum halepense*, *Setaria* sp. and broadleaf weeds like *Amaranthus* sp., *Portulaca* sp., *Ipomea* sp., *Chenopodium* album etc.

It is absorbed mainly through the root system and translocated for the leaves. It inhibits photosynthesis and interferes with nitrogen metabolism. It is used as pre-emergence treatment; provided adequate moisture is available in the soil to activate the herbicide.

It is strongly absorbed on soil and leaches slowly and more persistent in soils than monuron. Thus, the residue build up is high. The absorption is more in clay soil with organic matter content.

It is applied at 0.5 to 4 kg/ha in crop areas. 4 to 10 kg/ha for control of annual weeds and 15 to 30 kg/ha for perennials in non-cropped areas. It can be applied at 1 ppm as algaecide for treatment of ditches.

Fenuron (Urab): WP 50%

Fenuron is a non-selective herbicide, used for temporary sterilization in non-cropped areas and also as a brush control. It is used for killing deep-rooted perennials in moderate rainfall areas. It is also used to control weeds in citrus orchards. It is useful in crops like sugarcane, pineapple, and citrus. Absorption by the roots is the main mode of entry. It can be used only for spot treatment against perennial weeds in crop lands.

It is applied at 15 to 320 kg/ha and in granular form at 80 to 100 kg/ha. It is highly water soluble compound, easily leached in the soil and forms a good soil sterilant.

Fluometuron (Cotoran): WP 50%

Fluometuron is relatively non-selective, pre-emergence herbicide, usually applied to the soil surface before planting. It is also applied as post-emergence at very early stage of weed growth.

It controls most annual grass and broadleaved weeds under optimum moisture condition. The weeds controlled are *Echinochloa* sp., *Digitaria* sp., *Setaria* sp., *Eleusine* sp. and broadleaf weeds like *Ipomea* sp., *Chenopodium* sp. etc.

It is mostly used in crops like cotton as per-emergence. It can also be used in sorghum, sunflower, sugarcane, coffee, rubber, pineapple etc.; but found sensitive to cole crops, curcurbits, brinjal and sugarbeet.

It absorbs principally through roots. It persists in the soil for longer period, 8 to 20 weeks. It is also applied at 0.5 to 2 kg/ha for cotton and 2 to 3 kg/ha for sugarcane.

Monuron (Telvar): WP 50%

Monuron is a non-selective herbicide for general weed control in non-cropped areas. It is also used as selective pre-emergence herbicide in low doses (1 to 3 kg/ha) to control annual weed on tolerant crops like pineapple, sugarcane, spinach, citrus and grapes. The best result is obtained when applied to just emerged weeds. It should be applied far away from the fruit trees.

The grassy weeds controlled are: *Digitaria* sp., *Echonochloa crusgalli*, *Setaria* sp., *Sorghum halepense*; and broadleaved weeds like *Amaranthus*, *Portulaca*, *Chenopodium*, *Convolvulus* sp. etc.

It is readily absorbed through root system and is translocated upward in the xylem. It inhibits photosynthesis and interferes with nitrogen metabolism. At higher rates, phytotoxicity persists more than one season.

It is applied at 0.5 to 5 kg/ha; but as a soil sterilant at 10 to 15 kg/ha against perennial weeds. It is applied at 1 ppm as algaecide for treatment of drainage channels.

Alkoxy

Linuron (Lorox, Afalon): WP 50%

Linuron is a selective, pre-emergence herbicide which controls germinating and emerged grasses (upto 5 cm high) and broadleaved weeds (upto 15 cm high) in crops like soybean maize, sorghum, potatoes, cotton etc. It is also used for short-term control of annual weeds in non-cropped are: *Echinochloa crusgalli*, *Setaria* sp., *DIgitaria* sp., and broadleaf weeds like *Chenopodium album*, *Amaranthus* sp. *Portulaca oleracea* etc. This is usually spread on soil for effective weed control. Foliar spray of linuron can activate young weeds with suitable surfactants.

It is most rapidly absorbed through the root system and translocated upward in the xylem. The absorption increases on clay soil but phytotoxic effect decreases within 4 months. Its foliar absorption is greater than that of diuron, monurno and fenuron.

It is applied at 0.5 to 2 kg/ha in cropped area; in potato @ 1.25

kg; in pea @ 0.8 kg and beans @ 6.07 kg/ha, while in non-cropped areas @ 2 to 3 kg/ha. It persists in the soil up to 10 to 12 weeks.

Chlorbromuron (Maloran): WP 50%

Chlorbromuron is a selective, per-emergence herbicide which is equally effective through soil and foliage. It is broadcast or band treatment to the surface of the soil. It is also applied as post-emergence even the weeds reach 5 to 7 cm high. It can be applied in carrots, peas, maize, potato, soybeans, sorghum, sunflower and woody ornamentals; but the susceptible crops are sugarbeet, cole crops, cucurbits, tomato, okra and rice.

It is applied at 1.0 to 2.0 kg/ha.

Neburon (Bonus): WP 60%

It is a pre-emergence non-selective herbicide mostly used in lawns, turfs and woody rose nursery. It is more persistent in soil than Chloroxuron. It is applied at 2 kg/ha.

Unclassified

Methazole (Probe): WP 75%

Methazole is a herbicide which is applied as pre-emergence on the surface of the soil; as preplant on soybean sand as post-emergence on the tolerant crops such as cotton, sorghum and soybeans. Other crops which shows good selectivity are onion, grapes, citrus, wheat, and fruit orchards. The weeds controlled are: among broadleaf weeds – *Amaranthus*, *Ambrosia*, *Convolvulus*, *Euphorbia* sp., etc; and grassy weeds include *Digitaria*, *Echinochloa*, *Panicum* sp., *Poa annua* etc.

Probe does not move into young growing points, but mostly accumulates in older leaves. It is absorbed on the soil but does not persist. It is applied at 1.0 to 2.5 kg/ha as pre-emergence or post-emergence treatment.

Perfluidon (Destun)

Destun is a pre-emergence herbicide, applied to the soil surface for the control of nutsedge, grassy weeds and certain broad-leaved weeds. It is applied at 2 to 4 kg/ha.

C-288: Gr. 2.5%

C. 288 is a granular contact aquatic herbicide which is widely used in controlling of *Salvinia*, and aquatic floating weed which has become troublesome in wet land rice fields, channels etc. It also controls submerged weeds. However, the water level has to be brought down upto 5 to 10 cm before transplanting of rice, for its efficacy.

It is absorbed rapidly and kills the weeds without any harm to standing paddy crop. The field become completely weed free within a week. In wet land rice, the major weeds controlled are: *Salvinia molesta*, *Monochoria hastate* (pickerel), *Echinochloa crugalli* (barnyard grass).

It is applied 0.75 kg ai/ha (30 kg formulation/ha). It is slightly toxic and harmful to fish.

CHAPTER 9

Weed Control Technology and Training

Technology

Weed control technology is to be considered as a part of the production system, primarily to increase yield, to maintain soil fertility and labour utilization intensity during cropping season. The weed research thus has to play a unique role in an inter-disciplinary and integrated approach, to protect the crop from enormous losses by weeds, to secure the soil fertility, to permit seasonal engagement of available labours and to decrease the energy input to a minimum.

Changing of weed control practice from traditional manual or mechanical to herbicidal control must have a diverse impact on the whole crop production procedure. However, herbicide use in the developing countries has its own limitations due to high percentage of population engaged in agriculture as there is less opportunity for alternative highly paid rural employment. Even then, manual weeding is most expensive, item consuming and hence timely weeding not possible during peak cultivation period.

Weed control technology is inter-related to: rotational cropping system; cultivation method, spacing and time of planting; soil moisture availability and micro-climate etc. Hence the technology must be further; developed and made suitable for weed control in new rotational and cropping system; in soil conservation or water conservation programmes; so as to increase in yield per unit area. Other new problems to be tackled include:

Weed control in mixed-cropping situation; weed suppression for establishment of leguminous cover crops etc.

It is also necessary to evaluate appropriate weed management system that:

1. broad spectrum of weed control;
2. could be used before planting;

3. would reduce labour requirements;
4. would enhance crop production; and
5. would be based no easily understood, adoptable technology.

Training

The approach of weed problem should be considered as a whole system wherein the technology, training of technicians, economics, public information and sociological aspects of weed control will be involved. From the very beginning, weeds have been an integral part of agriculture; but the extent of loss due to weed infestation has not been properly assessed and their damage has not been considered so serious as caused by insects and diseases. It is realized in recent years that weed control is a pre-requisite for increasing production.

Herbicide use in small farms is often impractical due to lack of technical training and experience of the farmer. Hence, weed control which was so far not given must importance in the agriculture disciplines, is gaining momentum in the agricultural universities and other organizations. Lack of such training so far makes it difficult to maintain an adequate level of competence at various levels. Hence, there is an urgent need to train many workers and also educate the extension agencies properly and the farmers as well about herbicides and their proper uses. They should now the cost of herbicide application vis-à-vis manual or mechanical methods operation; periodicity and frequency of application for each type of weed; residual toxicity in the soil-affecting microbial flora, and also toxicity to host plant due to accidental exposure to the sprays etc.

Weed science training course should be comprehensive and practical field-oriented. This will include:

1. Study of weeds; crop losses due to weeds; prevention, cultural, mechanical and chemical control; absorption, translocation and selectivity of herbicides; properties of different herbicide groups; phytotoxicty and residual effect of herbicides; special problem on weed control in specific crops; non-selective weed control etc.
2. Publication of leaflets and bulletins, compilation of handbooks etc. for the benefit of the extension and developmental agencies and farming communities.
3. Farming should get information on herbicides from variety of sources including radio, television, news paper, company

agents and personal contacts. Like other plant protection chemicals, subsidies are essential for herbicide acceptance as the cost is comparatively high.

APPENDIX 1

Relative Mobility Classes of Herbicides in Soils in Ascending Order from Left to Right (After Helting and Turner 1968)

Immobile...Highly mobile

Mobility classes				
I	II	III	IV	V
Chloroxuron	Chlorprophan	Atrazine	Bromacil	Chloramben
Diquat	Dichlobenil	Alachlor (Lasso)	Dinoseb	Dalapon
Neburon	Diuron	Ametryne	Fenac	Dicamba
Nitralin	EPTC	Endothall	MCPA	TCA
Paraquat	Fenuron, Linuron	Flumeturon	Picloram	
Trifluralin	Molinate	Monuron	2,4-D	
	Propalin	Naptalam		
	Prometryne	Propachlor		
	Terbutryn	Prometone		
		Propham		
		Propazine		
		Simazine		
		Terbacil		
		2,4,5-T		

APPENDIX 2

Susceptibility of Crops on Certain Herbicides (After Hilton et al., WSSA 1974)

Herbicides	**Susceptible crops**
Barban	Oat, rye
Chlorbromuron	Sugarbeet, cole crops, cucurbits, tomato, okra, rice
Chloroxuron	Sugarbeet, cole crops
2,4-D (amine)	Dicot plants
2,4-DEP (Falone)	Cotton, tobacco, tomato, onion, grapes
Dicamba	Soybean, beans, small-seeded legumes, ornamentals, vegetables
Dinoseb	Cruciferous crops
Flumeturon	Sugarbeet, cole crops, cucurbits, brinjal
Fluorodifen	Sugarbeet, cucurbits tomato, alfalfa
Metribuzin	Sugarbeet, cole crops, cruciferous crops, onion, pea, sunflower, sweet-potato, cotton and tobacco
Naptalam	Sugarbeet, tomato, spinach
Picloram	Broadleaved plants except cruciferous crops
Propazine	Sugarbeet, vegetables
Simazine	Sugarbeet, vegetables, tobacco
TCA	Cereals and millets

APPENDIX 3

Organic Herbicides

Common name (trade name)	Molecular formula	Mol. wt.
Organic herbicides		
Aliphatics		
Dalapon (Dowpon)	$CH_3C.Cl_2COOH$	165
TCA (Varitox)	$Cl_3C, COOH$	163.5
Anilides and Amides		
Alachlor (Lasso)	$C_{14}H_{20}ClNO_2$	269.5
Butachlor (Machete)	$C_{17}H_{26}ClNO_3$	312
Propachlor (Ramrod)	$C_{11}H_{14}ClNO$	211.5
CDAA (Randox)	$C_8H_{12}ClNO$	173
Propanil (Stam F-34)	$C_9H_9Cl_2NO$	218
Naptalam (Alanap)	$C_{18}H_{13}NO_3$	29
Pronamide (Kerb)	$C_{12}H_{12}Cl_2NO$	256
Anilines & Nitrophenols		
Dinitramine (Cobex)	$C_{11}H_{13}F_3N_4O_4$	322
Nitralin (Planavin)	$C_{13}H_{19}N_3O_6S$	345
Trifluralin (Treflan)	$C_{13}H_{16}F_3N_3I_4$	335
Dinoseb (Pre-emerge)	$C_{10}H_{12}N_2O_5$	240
PCP (Durotox)	HCl_5O	266
Nitrofen (TOK)	$C_{12}H_7Cl_2NO_3$	284
Fluorodifen (Soyex)	$C_{13}H_7F_3N_2O_5$	32
Arsenicals		
MAA (Ansar)	CH_5ASO_3	
DSMA (Chipco, carb kleen)	$CH_3ASO_4(ONa)_2$	144
MSMA (Daconate)	$CH_3AS.OH(ONa)$	162
Benzoics		
Chloramben (Amiben)	$C_7H_5Cl_2NO_2$	206
Dicamba (Banvel)	$C_8H_6Cl_2O_3$	221
Fenac (Fenac)	$C_8H_5Cl_3O_2$	239.5
Carbamates		

Common name (trade name)	Molecular formula	Mol. wt.
Asulum (Asulox)	$C_6H_{16}N_2O_4S$	230
Barban (Carbyne)	$C_{11}H_9Cl_2NO_2$	258
Chlorpropham (CIPC)	$C_{10}H_{12}ClNO_3$	213.5
Propham (IPC)	$C_{10}H_{13}NO_2$	179
Thio-carbamate		
Benthiocarb (Saturn)	$C_{12}H_{16}ClNOS$	257.5
EPTC (Eptam)	$C_9H_{19}NOS$	189
Diallate (Avadex)	$C_{10}H_{17}Cl_2NOS$	270
Triallate (Avadex BW)	$C_{10}H_{16}Cl_2NOS$	304.5
CDEC (Vegadex)	$C_8H_{14}ClNS_2$	224
Molinate (Ordram)	$C_9H_{19}NOS$	187
Glyphosate (Round up)	$C_3H_8NO_5P$	169
Helerocyclic compounds		
Bipyridylium		
Diquat (Reglone)	$C_{12}H_{12}N_2X_2$	344
Paraquat (Gramoxone)	$C_{12}H_{14}N_2Cl_2$	257
Pyridines		
Picloram (Tordon)	$C_6H_3Cl_3N_2O_2$	241.5
Pyridazines		
MH(Retard)	$C_4H_4N_2O_2$	112.1
Uracils		
Bromacil (Hyvar)	$C_9H_{13}BrN_2O_2$	261
Terbacil (Sinbar)	$C_9H_{13}ClN_2O_2$	
Bentazon (Besagram)	$C_{10}H_{12}N_2O_3S$	240
Triaziens		
Atrazine (Atrataf)	$C_8H_{14}ClN_5$	215.5
Propazine (Gesamil)	$C_9H_{15}N_5Cl$	229.5
Ametryne (Gesapax)	$C_9H_{17}N_5S$	227.5
Prometone (PRamitol)	$C_{10}H_{19}N_5O$	225
Prometryne (Gesagard)	$C_{10}H_{19}N_5S$	241
Simazine (Gesatop)	$C_7H_{12}ClN_5$	201.5
Metribuzin (Sencor)	$C_8H_{14}N_4OS$	214
Terbutryn (Igran)	$C_{10}H_{19}N_5S$	241
Misc. Heterocyclic compounds		
Amitrole (Weedazol)	$C_2H_4N_4$	84
Endothall (Endothall)	$C_8H_{10}O_5$	186

Common name (trade name)	Molecular formula	Mol. wt.
Oxadiazon (Ronstar)	$C_{15}H_{18}O_3N_2Cl_2$	345
Pyrazon (Pyramin)	$C_{10}H_8ClN_3O$	221.5
Hormone type herbicides		
2,4-D (Fernoxone Weedone)		
2,4-D	2,4,D-Amine	2,4-D ester
$C_8H_6Cl_2O_3$	$C_{10}H_{13}Cl_2NO_3$ $C_{14}H_{18}Cl_2O_4$	221
2,4,5-T (Brush killer)	$C_8H_5Cl_3O_3$	255.5
MCPA (Chiptox)	$C_9H_9ClO_3$	200.5
Dichlorprop (2,4-DP)	$C_9H_8Cl_2O_3$	235
Mecoprop (MCPP)	$C_{10}H_{11}ClO_3$	213.5
Silvex (Fenoprop)	$C_9H_7Cl_3O_3$	267.5
2,4-D-B (Butaxone)	$C_{10}H_{10}Cl_2O_3$	249
MCPB (Tropotox)	$C_{11}H_{13}ClO_3$	228.5
2,4-DEP (Falone)	$C_{24}H_{21}Cl_6O_6P$	649
Dichlobenil (Casoran)	$C_7H_3Cl_2N$	172
Ioxynil (Actril)	$C_7H_3I_2NO$	371
Chloroxuron (Tenoran)	$C_{15}H_{15}ClN_2O_2$	291
Fenuron (Urab)	$C_{11}H_{13}Cl_3N_2O_3$	328
Diuron (Karmex)	$C_9H_{10}Cl_2N_2O$	233
Flumeturon (Cotoran)	$C_{10}H_{11}F_3N_2O$	232
Monurno (Telvac)	$C_9H_{11}ClN_2O$	362
Linuron (Lorox)	$C_9H_{10}Cl_2N_2O_2$	249
Chlorbromuron (Maloran)	$C_9H_{10}BrClN_2O_2$	294

APPENDIX 4

List of Herbicides, Rate and Mode of Treatment

Common name	Trade name	Conc.	Mode of treatment	Rate kg/l per ha.
Acrolein	Aqualin	EC 92%	-	50 ppm
Alachlor	Lasso	EC 48%	Pre-	2-3
		Gr 10%		
AMS	Ammate	WP 68%		
Ametryn	Evik	WP 80%	Pre/early post	2-3
Aminotriazole	Weeder	WP 50%		
Amitrole	Weedazol, amizole	WP 50%	Pre, Post	2-5
Ammonium sulphamate	Silvicide	CR 95%		
Asulam	Asulox	EC 40%	Pre-	2-4
Atrazine	Atrataf, actrex	WP 50%	Pre-	0.5-2
Barban	Carbine	EC 25%	Post-	0.5-1
Benefin	Emblem	EC 65%		
Bentazon	Basagran	WP 48%	Pre-	0.5-1.5
Benthiocarb	Saturn, Bolero	EC 50%		
	Bolero	Gr. 10%		
Bifenex	Modown	WP 80%		
Borate	Hibour, Monobor, Poybor	WP 94%	Soil Pre/post	
Bromacil	Hyvar-X, Urox-H	WP 80%	Siol Pre/post	2-3
Bromoxynil	Brominal, buctril, Nu-Lawn	EC 35%	Soil, Post	0.2-0.5
Butachlor	Machete	EC 50% Gr. 5%	Pre-	
Cacodylic acid	Phytar-560, Kilz-All, Red-E-cate 35	EC 67%		
CDAA	Randox	GR 20%	Pre-	3-8

Common name	Trade name	Conc.	Mode of treatment	Rate kg/l per ha.
CDEC	Vegadex	GR 20%	Pre-	2-4
C-288		GR 2.5%	Post	0.75 ai 30 kg
Chloramben	Amiben	EC 24% Gr. 4%	Pre-	2-4
Chlorbromuron	Maloran, Bromex	WP 50%	Pre-	1-2
Chloroxuron	Tenoran,	EC 40%	Pre-	1-3
Chlorpropham	Furloe	EC 40%	Pre-	1-3
Copper sulphate	Cuprose	S 100%		
Cyanazine	Bladex	WP 50%		
2,4-D	Aqua-kleen,	SP 80%	Post	0.5-2
2,4-D (ethyl ester)	Weedone			
2,4-D (amine)	Ded-weed 40, DMA-4, Emulsavert			
2,4-D (ester)	Weed killer 646			
2,4-D (sodium)	Fernoxone, Weedox			
2,4-D (dimethyl amine)	Weeder-96			
2,4-D (special formulation)	Decamine			
Dalapon-sodium	Dowpon, Tafapon, Radaon	WP 85%	Pre/post	
2,4-db	BUTOXONE, Butyrac Ester	EC 24%	Post	2-5
DCPA	Dacthal	WP 75%	Pre-	
Desmetryne	Semeron	WP 25%		
2,4-DEP	Felone	WP 25%	Pre-	1-3
2,4-DES	Sesone	WP 98%	Pre-	2-5
Diallate	Avadex	EC 40%	Pre-	2-4
Dicamba	Banvel	EC 50%	Pre/post	0.5-3
Dichlobenil	Casoran	WP 45%	Pre-	1-3
Dichlormate	Sirmate	EC 80%	Pre-	1-2
		Gr. 5%		
Dichlorprop	Weedone, 2,4-DP	EC 50%	Post	7-9
Dinitramine	Cobex	EC 25%	Pre-	0.5
DNOC	Sinox	EC 46%		
Dinoseb (DNBP)	Dow weed killer, Sinox	EC 48%	Post	0.5-1.5

Common name	Trade name	Conc.	Mode of treatment	Rate kg/l per ha.
	general Premerge			
Diphenamid	Dymid, Enide	WP 80%	(Soil-steri)	
Diuron	Karmex	WP 80%	Soil, Pre-	0.5-4
Diquat-	Reglone	EC 20%	Post	0.5-1.5
DSMA	Chipco, carb			
	Kleen	EC 35%	Pre/post	2-3
2,4,5-T	Brush-killer Tinoxol	EC 48%	Post	1-3
2,4-5-T (amine)	Veon-245			
2,3,6-TBA	Fen-All, Trysben-200		Pre/post	1-3
Endothall	Hydrothol-191	EC 55%	Pre/post	1-5 ppmw
EPTC	Eptam	EC 75%	Pre-	2-4
		Gr. 10%		
Fenac	Fenac	WP 40%	Pre	2-5
Fenuron	Dyber, Urab	WP 50%	Soil, Pre	5-10
Fluchloralin	Basalin	EC 48%	Pre-	1-2
Fluometuron	Cotoran, Lanex	WP 50%	Soil, Pre	0.5-2
Flurodifen	Soyex	EC 30%	Pre-	3-5
Glyphosate	Round up	EC 41%	Post	1-2
Ioxynil	Actril	EC 25%	Pre-	0.5-1
Iron sulphate	Antimos	WP 96%		
Linuron	Afalon, Lorax	WP 50%	Pre-	0.5-2
MAA	Ansar, Weed-E-Rad	EC 35%	Post	2-3
MCPA	Chiptox	WP 75%	Post	0.25-2
MCPB	Cantrol, Thistrol, Tropotox	EC 40%	Post	0.5-1.5
Mecoprop	MCPP, Lornox	EC 50%	Post	1.5-2.5
Methazole	Probe	WP 75%	Pre/Post	1-2.5
Metobromuron	Patoran	WP 50%		
Metribuzin	Lexone, Sencor	WP 70%	Pre-	0.25-1
		Gr. 7%		
MH	Retard	EC 30%	Post	2.5-5
Molinate	Ordrum	EC 72%	Pre-	2-3
Monuron	Telvar	WP 50%	Pre-	1
MSMA	Ansar 529, deconate, Merge-823, Bueno	EC 35%	Post	2-3
Naptalam	Alanap, NPA-3	EC 24%	Pre-	2-4

Common name	Trade name	Conc.	Mode of treatment	Rate kg/l per ha.
Neburon	Bonus	WP 60%		
Nitralin	Planavin	WP 75%	Pre-	1
Nitrofen	Tok-E-25	EC 25%	Pre/Post	2-5
Oxadiazon	Ronstar	EC 25%	Early Post	0.75-4
Paraquat	Gramoxone	EC 10%	Post	0.5-1
PCP	Durotox		Pre/Post	2-5
Penoxalin	Prowl, Stomp			
Perfluidon	Destun		Pre-	2-4
Picloram	Tordon, Amdon	EC 25%	Pre/Post	2-4
Prometone	Pramitol 25E	EC 25%	Pre/post	6-10
Prometryne	Gasagard	WP 50%	Pre-	1.5-3
Prometyn	Caparol			
Pronamide	Kerb	WP 50%	Pre-	0.5-2
Propachlor	Ramrod	WP 65%	Pre-	
Propanil	Rogue, Stam F-34, Surcopur	EC 35%	Post	
Propazine	Milogard	WP 50%	Pre-	0.5-2
Propham	Chem-Hoe	WP 50%	Pre-	1-4
Pyrazon	Pyramin	WP 45%	Pre/Post	2-4
Silvex (Fenoprop)	Kuron		Post	1
Simazine	Princep	WP 50% Gr. 20%	Pre-	1-3
sodium chlorate	Atlacide	WP 10%		
TCA	Dow Sodium TCA	WP 90%	Soil, Pre	
Terbacil	Sinbar	WP 80%	Soil, Pre	1-3
Terbutryn	Igran	WP 80%	Pre/Post	1.5-2
Triallate	Avadex BW, Far-Go	EC 40%	Soil, Pre/Post	1.5-2.2
Trifluralin	Treflan	EC 48%	Soil, Pre	0.5-1.5
Vernolate	Surpass	EC 72% Gr. 10%		

APPENDIX 5

Herbicide Combinations

	Combination	Trade name
1.	Atrazine + Propachlor	Astram
2.	Atrazine + Prometone	Atratol 80W
3.	Atrazine + Simazine	Pintex
4.	Atrazine + Amitrole + Fenac	Fenamine
5.	Atrazine + Chlorate + Borate	Atratol 80W
6.	Borate + Bromacil	Borasen
7.	Borate + Diuron	Monobor-chlorate
8.	Borate + Monuron	Ureabor
9.	Borate + Sodium chlorate	Chlorax-40
10.	Borate + 2,3,6-TBA	Benzabor
11.	CDAA + TCBC	Randox T
12.	Dalapon + TCA	Dowpon
13.	Diuron + Bromoxuron	Brute
14.	Diuron + Bromacil	Krovar
15.	2,4-D + Borax	DB granular
16.	2,4-D + Fenac	Banevel, Brush Buster
17.	2,4-D + Dicamba	Banevel, Brush Bluster
		Banvel-K, Weedone Super D.
18.	2,4-D + Dichlorprop	Bush killer 170
19.	2,4-D + EPTC	Knoxweed
20.	2,4-D + MCPP	Chipco Turf Kleen
21.	2,4-D + Mecoprop	Mecopar
22.	2,4-D + MSMA	Mad
23.	2,4-D + Picloram	Amdon 101, Tordon 101, Tordon 212
24.	2,4-D + 2,4-DP	Envert-171, weedone-170
25.	2,4-D + 2,4,5-T	Amine 2D/2T, ded-weed LV 33, Dinoxol, Envert 171, Esteron, Emulsamine BK, Emulsavert 248, Lo-Vol 2D/2T, Brushkiller 64, Weeder-Amine-BK, Weedone BK 64

	Combination	Trade name
26.	2,4-D + 2,4,5-T + Dicamba	Banvel Brush Buster-111
27.	2,4-D + DSMA + Bromacil	Wood broom
28.	2,4-D + Surfactant	Weed-E-Rad 360, Ansar, DSMA liquid
29.	Endothall + Silvex	Aquathol plus
30.	Endothall + Ammonium sulphate	Des-i-cate
31.	Linuron + Amitrole	Kleer-Lot
32.	Linuron + Chloramben	Amilon
33.	Linuron + Propachlor	Londax
34.	MCPA + Bromoxynil	Brominal plus, Bronate
35.	MCPA + Dicamba	Mno Dak
36.	MSMA + Surfactant	Ansar 529 HC, Weed Hoe-108
37.	MSMA + Cacodylic acid	Chex-mate
38.	Naptalam + Chlorpropham	Alanap plus, Mor-corn, Solo
39.	Naptalam + Dinitro	Kleenkrop
40.	Trifluralin + Diphenamid	Trefmid

APPENDIX 6

Selective Bibliography

Plenty of references on specific weeds and weedicides are available and it is difficult to cite all of them. Hence only selective ones, reference books and manuals are listed. For details, standard books, modern weed science etc. may be referred.

1. Ahrens, J. F. 1972. Granular herbicides, Weeds Today. 3(2): 12-13.
2. Anonymous, 1953. Control of Orobanche on Tobacco. Plant Prot. Bull. 5: 27.
3. Anonymous, 1964. Weed control in vegetables. Farm Tech. Bull. Univ. Calif. March 13-17 pp.
4. Anonymous, 1964. Report of the Terminology Committee, Weed Society of American Weeds 12(4): 3p.
5. Anonymous, 1967. Chemicals weed control in rubber and oil palms. Agri. News BASF No. 8-12 pp.
6. Anonymous, 1969. Proceedings of the second Asia-Pacific Weed Control Interchange, IRRI, Philippines, June 16-20.
7. Anonymous, 1973. 3rd Inter. Symp. on Biological control of weeds. Montpellier, France. PANS 20(1): 111-114.
8. Ashton, F. H. and Crafts, A. S. 1973. Mode of action of herbicides, John Wiley & Sons, NY. 504 pp.
9. Beilin, I. G. 1967. Control of dodder and broomrape. Kolos Moskva 38 pp. (Weed abstr. 1303: 18(3): 1969.
10. Bharadwaj, R. B. L. and Verma, R. D. 1968. Seasonal development of nut grass under Delhi conditions. Ind. J. Agri. 33(6): 950-57.
11. Bindra, O. S. and Singh, H. 1971. Pesticide application equipments. Oxford & IBH Pub. Co., New Delhi. 397 pp.
12. Brian, R. C. 1964. The metabolism of herbicides. Weed Res. 4(2): 105-107.
13. Butler, J. D. and Slife, F. W. 1970. Lawn weeds: Identification and control. Univ. Illin. Coll. Agri. Boop Extn. Serv. Circ. 873. 28.
14. Cardanas, J. Reyes, C. E., Doll, J. D. 1972. Tropical weeds Vol. II, Colombia Inter Platn Prot. Centre, Oxegon State Univ.
15. Chakravarti, A. K. 1957. Weed flora of paddy fields of West Bengal. Ind. Agriculturist 1(2): 19-26.
16. Crafts, A. S. and Drever, H. 1960. Experiments with herbicides in soils. Weeds 8(1): 12-18.
17. Crafts, A. S. and Robbins, W. W. 1962. Weed control. McGraw Hill Book & Co., New York N.Y. 10036, 660 p.
18. Danks, M. L., Fletcher, J. S. and Rice, E. L. 1975a. Influence of ferulic acid on mineral depletion and uptake of 86Rb by paul's Scarlet rose cell-

suspension cultures; Am. J. Bot. 62: 749-755.
19. Dodd, A. P. 1957. Recent development in Biological control of Lantana, Minutes Ent. Soc., Queensland, Nov.
20. Ennis, W. B. Jr. 1964. Selective toxicity in herbicides. Weed Res. 4: 93-104.
21. Frear, Donald, E. H. 1955. Chemistry of Pesticides, D. Van Nostrand Co., Inc. New York.
22. Gupta, O. P. 1973. Aquatic weed control. World Crops 25(4): 182-89.
23. Gupta, O. P. and Lamba, P. S. 1984. Modern weed science. Today and Tomorrow, New Delhi. 421 p.
24. Hag, A. 1955. Weed flora of paddy fields and its control in U.P. Science and Culture 21: 277-78.
25. Harley, K. L. S. 1971. Biological control of Lantana, PANS 17(4): 433-37.
26. Helling, C. S., Turner, B. C. 1955. Pesticide mobility determination in this layer chlomotography. Sci. 162: 526-563.
27. Hilton, J. L., Bovey, R. W., Hull, H. M., Mullison, W. R., Talbet, R. E. 1974. Weed Science society of America. Herbicide Hand Book W.F. Humphrey Press Inc. N.Y. 368 p.
28. Hutchinson, J. 1926. The families of flowering plants (2 vol.) Macmillan & Co. Ltd., London 792 p.
29. Joshi, N. C. 1973. Some problems and progress of weed control in India. 1948-72. 4th Asia-Pacific Weed Sci. Soc. Conf. Newzealand. 170-76.
30. Joshi, N. C. 1974. Manual of weed control, Research Co., Pub. Delhi.
31. Kadambi, K. 1954. Loranthus Control. Ind. Forester. 80: 493-94.
32. Kaul, R. N. 1957. Studies on weed-crop competition. Allahabad Farmer 15(2): 103-15.
33. Kasasian, L. 1971. 2nd Inter. Symp on Biological control of weeds. Conf. Rep. PANS 17(4): 517-19.
34. Khajwria, R. R. 1967. Plants that may do mischief in pastures. Ind. Fmg. Jan. 2.
35. Khuspe, V. S., Subbaiah, R., Mande, J. V. 1982. A compendium of Indian Weed Science Research (2 Vol.). Metropolitian Book Co., New Delhi. 871p + 564 p.
36. Klingman, G. C. 1961. Weed control as a Science, John Wiley & Sons, Inc. New York. 421 p.
37. Lange, A. J., Cardenas and Cruis, R. 1973. Crop losses caused by weeds, Weeds Today 4(1): 11-14.
38. Little, E. C. S. 1967. Progress report on transpiration of some tropical weeds PANS(c) 13(2): 127-131.
39. Mandal, R. C. 1977. Effect of newer herbicides on weeds in dry and wet-land rice in Fiji. In Proc. ISWS/APAU Weeds Sci. Conf. Hyderabad. P. 176.
40. Mani, V. S., Gautam and Chakravarthy, T. K. 1968. Losses in crop yields due to weed growth. PANS(c) 14: 142-158.
41. Mani, V. S. 1975. Nutrient drain weed growth in crop fields. Fert. New. 20(2): 21-27.
42. Matsunaka, S. 1969. Activation and inactivation of herbicides by higher plants. Residue Rev. 25: 45-58.
43. Muzik, T. J. 1970. Weed biology and control. McGraw Hill Bok & Co., NY 273 p.
44. Mukhopadhaya, S. K. 1974. Increasing fertilizer efficiency through weed

control. Fert News. 19(12): 56-58.
45. Penner, D. 1974. The effectiveness of pesticide interactions. Weeds Today 5(3): 13-14.
46. Rao, V. P. 1969. Possibilities of biological control of aquatic weeds in India. ECAFE Water Resources J. 40-50 pp.
47. Rice, Elroy, L. 1984. Allelopathy (2nd Edition), Academic Press, Inc. London.
48. Robinson, E. L. 1976. Effect of weed species and placement on seed cotton yields. Weed Sci. 24: 353-55.
49. Sharma, D. K. 1960. Control of Orobanche (broom rape) and have better yields. Food Fmg. 12: 28.
50. Sharma, V. S. 1977. Weed management in Tea fields in South India, Ind. J. Weed Sci. 9(2): 146-153.
51. Singh, M., Pandy, R. K., Dutta, T. R. 1971. Weed problem and control in grassland and fodder crops in India. Ind. J. Weed Sci. 3(1): 45-50.
52. Sinha, T. D. and Thakur, C. 1967. Control of nutgrass weed by cultivation. Ind. J. Agron. 12(1): 121-125.
53. Smith, H. P. 1964. Farm machinery and equipment, McGraw-Hill Book Co., NY 134-165 and 209-235.
54. Suryawansi, S. M. 1980. Efficacy of farm chemicals to control Parthenium hysterophorus. ISWS/OUAT Weed Sci. Conf. Bhubaneswar.
55. Tadulingam, C. 1985. A Handbook of some South Indian Weeds. Govt. Press. Madras. 488 p.
56. Tanus, R. S., Gastro, M. D. V. and Vieitex, E. 1973. Growth substances isolated from tubers of *Cyperus escularitus* var. *aureus*. Physiol. Plant. 28: 195-200.
57. Thakur, C. 1957. Weeds in Indian agriculture. Motilal Banarsi Das Publishing, Bankipur, Patna.
58. Thakur, C. 1984. Weed Sciences. Metropolitan Book Co., New Delhi. 495 p.
59. Thomas, K. M., Narasimhawami, R. L. 1952. Weed control in coffee plantation. Plrs chron. 47: 606-7.
60. Vengris, J. M., Drake, W. G. Colby and Bart, J. 1953. Chemical composition of weeds and accompanying crop plants. Agron. J. 45(5): 213-18.
61. Wain, R. L. 1955. A new approach to selective weed control. Ann. Appl. Biol. 42: 151-157.
62. Weldon, L. W., Blackburn, R. D. and Harrison, D. S. 1973. Common aquatic weeds. Dover Publications Inc. New York. 43 p.
63. Wiese, A. F. and Davis, R. G. 1964. Herbicide movement in soil with various amounts of water. Weeds 12(1): 101-103.
64. Wiese, A. F. and Vandiver, C. W. 1970. Soil moisture effects on competitive ability of weeds. Weed Sci. 18(4): 518-19.

APPENDIX 7

Address of Chemical Companies and Source of Herbicides, Application Equipments and Weeding Implements

Firms	Herbicides
Agromore Pvt. Ltd., Mysore Road, Bangalore – 26	Mostly 2,4-D formulations; 2,4-D ethyl ester-Agrodone conc. 48 (34%, 18% WP 4%, Gr; Agrodar 96 (dimethyl amine) 58%, Agrodex diuron 80%.
Alkali & Chemical Crop. of India 19, Walchand Hirachand Marg, Bombay -1 or Hamilton House, Cannaught Place, New Delhi-1	Grammoxone, Diquat, Dalapon, Fernoxone
Allied Chemical Crop. PB No. 1057 R, Morristown New Jersey 07 960 USA	Fenuron, TCA, Monuron, Sodium
Amchem Products Inc. Ambler, Pennsylvania-19002 USA	Amitrol, Bromoxynil, Chloramben, 2,4-D, 2,4-DB, MCPA, Dichlorprop, Fenac, MAA, DSMA, MSMA, Naptalam, Silvex, 2,4,5-T
Ansul Co., Stanton Street, Winsconsin Winsconsin-54143 USA	TBA, MAA, DSMA, MSMA, MH
Basf India Ltd. May Baker House, PB 19108 Bombay-25 or Basant Vihar, New Delhi – 57	Bentazon, Fluchloralin, Dinoseb, Pyrazon
Bayer (India) Ltd. 82, Veer Nariman Road, Bombay-20 or 56, Janpath New Delhi – 1	Methabenzthirazuron, Metribuzin
Bharat Pulversing Mill Pvt. Ltd. 14, Queens Road Bombay-20	MCPA,2,4,-D ester (Bladex A,B,C,F & G), Fluometuron

Firms	Herbicides
Boots Pure Drug Co. Ltd. Dr. Annie Besant Road, Bombay-1	Mecoprop, Bentozolin
Chipman Div. of Rhodia Inc. 120 Jersey Avenue, New Jersey-08903 USA	Asulum, Oxadiazon, Bromoxynil, 2,4-DB, Diquat, MCPA, MCPB
Cynamid India Ltd. Nylock House, PB 9109, Baombay-25 or Bhandari House, New Delhi-2	Pendimethalin
Dow Chemical International Ltd. Kasturba Gandhi Marg, New Delhi-1	2,4-D Dalapoin, Diuron, Picloram Silvex, 2,4,5-T, TCA, Dinoseb.
Du-Pont de Nemours & Co. Wilmington Delawart-19898	AMS, Bromacil, Diuron, Monuron
Gulf Crop Protection Products 9009W, 67th Street Merriam Kansas-66202	Barban, Benzadox, Cyprazine
Hindustan CIBA-Giegy Ltd., Khetan Bhawan, 198, J. Tata Road Bombay – 400 020	Prometone, Ametryne, Prometryne, Atrazine, Chlorbromuron, Chloroxuron, Flumeturon, Fluorodofem Isoproturon, Terbutryn, Profluralin, Propazine, Simazine, Terbuthylazine
Hoechst Pharmaceuticals Ltd. 193, Backbay Reclamation, PB No. 11123 Bombay-21	Arelon (Iroproturon) 75% Afalon 50% (Linuron) Arazon 30% (Anilofos)
Illi Lilly & Company PB No. 708, Greenfield, Indiana – 46140	Benefin, Diphenamid, Iso-propalin, Trifluralin
Imperial Chemical Industries (ICI) 18, Strand Raod, Calcutta – 1	Diquat, Paraquat
Imkemex India Ltd. Crescent House, Ballard Estate, Bombay-1	Fernoxone
Indofil Chemical Ltd. Dr. E. Moses Road, PB 6280 Bombay-1	Oxyfluorfen, TOK-E 35, Propanil
Kumai Chemical Co., 6-2 Ohtemachi 2-Chome,	Benthiocarb

Firms	Herbicides
Chiyoda-ku Tokyo, Japan	
May & Baker Pvt. Ltd. Sir Phirozshah Mehta Road, Bombay – 1	Ioxynil, MCPB, TCAQ, Spontox 20% EC, Bladex-I
Merck AG Darmstadt, West Germany	Chloflurenol, Flurenol
Monsanto Chemicals (India) Pvt. Ltd. Wakefield Hosue, 11 Sprott Road, PB 344A, Bombay -1 or 3/8 Asaf Ali Road, New Delhi – or 157 Lingh chetty Street, Madras -1	MAA, MSMA, DSMA, CDAA, CDEC, Butachlor, Alachlor, Propachlor, Diallate
National Organic Chemical Indus. Ltd. 16, Parliament Street New Delhi -1	Isoproturon, Cynazine
Pacific Chemicals Hindustan Ltd. 18/20 K.G. Road New Delhi-1	Dalapon
Pennwalt Corporation 1713 South California Avenue, Monrovia, California California-91016 USA	Arsenic acid, Endothall, Sodium chlorate
Philipps Duphar B.V. PB 7006, Amsterdam-7, Netherlands	Dichlobenil
Phone-Poulenu Agro-clinic May Baker House, Worli Baombay-25	Isoproturon, Asullum, Terbacil, Oxadiazon
PPG Industries Inc. One Gateway Center, Pennsylvania – 15222, USA	Chlorpropham, Propham, Potassium azide, Sodium azide
Pesticides India Ltd. 9/122, Motibagh, Agra	Benthiocarb, Isoproturon
Rallis India Ltd. Ravelin Street, Bombay – 400 001	Atrazine, Simazine, Dalapon
Rahm & Haas Company Philadelhpia, Pennsylvania – 19105, USA	Nitrofen, Pronamide, Propalin
Sandox (India) Ltd. Dr. Annie Besant Road, PB 6596, Worli	Metoxuron

Firms	Herbicides
Bombay-18	
Shell Chemical Company 2401 Crow Canyon Road, California – 94583, USA	Acrolin, Nitralin, cyanazine
Stanffer Chemicals B-8 Defence Colony, New Delhi – 3	Eptam, EPTC, Molinate, Vernoalte, Cycolate
Tata-Fison Industries Ltd. Dalal Street, Bombay – 1	Atrazine, Simazine
Thomson Hayward Chemical Co., PB 2383, Kansas City – 66110	Dichlobenil
UD Borax & Chemical Corporation 412, Crescent Way, California – 92801, USA	Borate, Borax, Bromacil, Chlorfluranol, Dinitramine
Velsicol Corporation C-32, Panchsheel Enclave, New Delhi – 17	Dicamba

Herbicide Application Equipments: Suppliers

1. American Spring & Pressing Works Pvt. Ltd., Malnad, Bombay
2. Amil Agri-Horticultural Co (P) Ltd., 18, Chandra Puri, Ghaziabad – 201 001
3. Diamond Equipment Company, 46, Sowripalayam Road, Coimbatore -4.
4. Indo-German Plantation Machinery Ltd., Bangalore -12
5. Saw Wallace & Co. Ltd., 4, Bankshall Street, Calcutta – 1
6. Sigma Steel Industries, A-2, Industrial Estate, Ludhiana – 3

Weeding Implements: Suppliers

1. BECO Engineering Co. Ltd., C.T. Road, Batala, Punjab
2. Cossul & Co., Kanpur
3. Escorts Ltd., Faridabad
4. Kirloskar Bros. Ltd., Kirloskarwadi, Sangli, Maharashtra
5. Kumaon Nursery, Ramnagar, Nainital (U.P.) – Horticultural Implements.
6. Tractor & Farm Equipments, Madras
7. Voltas Ltd., 19, Graham Road, Ballard Estate, Bombay.

APPENDIX 8

World's Worse Weeds

Weed ranking	Botanical name	Common name	Type of weed	Annual / perennial	No. of crops affects
1.	*Cyperus rotundus*	Nut grass	Sedge	Perennial	52
2.	*Cynodon dactylon*	Bermuda grass	Grass	Perennial	40
3.	*Echinochloa crusgalli*	Barnyard grass	Grass	Annual	36
4.	*Echinochloa colonum*	Jungle rice	Grass	Annual	35
5.	*Eleusine indica*	Goose grass	Grass	Annual	46
6.	*Sorghum halepense*	Johnson grass	Grass	Perennial	30
7.	*Imperata cylindrical*	Johnson grass	Grass	Perennial	35
8.	*Eichhornia crassipes*	Water hyacinth	Herbaceous	Perennial	
9.	*Portulaca oleracea*	Common perslane	Herbaceous	Annual	45
10.	*Chenopodium album*	Lambs quarters	Herbaceous	Annual	40
11.	*Digitaria sanguinalis*	Crab grass	Grass	Annual	33
12.	*Convolvulus arvensis*	Field bindweed	Herbaceous	Perennial	32
13.	*Avena fatua*	Wild oat	Gass	Annual	30
14.	*Amaranthus hybridus*		Herbaceous	Annual	27
15.	*Amaranthus spinosus*	Spiny pigweed	Herbaceous	Annual	28
16.	*Cyperus esculantus*	Yellow nutsedge	Sedge	Perennial	21

Weed ranking	Botanical name	Common name	Type of weed	Annual / perennial	No. of crops affects
17.	*Paspalum conjugatum*	Sour grass	Grass	Perennial	25
18.	*Rottaboellia exaltata*	Raoul grass	Grass	Annual	18

After Holm *et al.* 1977 "The World Worst Weeds"

APPENDIX 9

Terms and Abbreviations

Achene. Simple, one-seeded, indehiscent fruit, seed attached to ovary wall.
Acid equivalent (ae). Active ingredient content of a formulation.
Acropetal. Youngest organ nearest to the apex.
Active ingredient (ai). A chemical formulation that is directly responsible for herbicidal effects.
Adventitious bud. Bud arising anywhere on the plant except leaf axils.
Alternate. One leaf at a node.
At emergence. During visible emerging phase of specified crop or weed where leaf joins the stem.
Axil. Where leaf joins the stem.
Basal treatment. Applied on the ground encircling the stem.
Berry. Fleshy fruit.
Biennial. Life-cycle completes within two years.
Bisexual. Containing both fertile stamens and carpels with ovules.
Bract. Scale-like leaves of a flower cluster.
Brush control. Control of weedy plant, clumps, shrubs etc.
Bulb. Underground fleshy stem with scale leaves.
Calyx. Outer whorl of floral parts.
Capsule. Dry dehiscent fruit with two or more carpels.
Carpel. A floral leaf bearing ovule.
Carrier. A liquid or solid substance used to dilute herbicide during its application.
Chlorophyll. Green colouring matter within the cells of plants.
Chlorosis. Loss of green colour in foliage, followed by yellowing of the tissue.
Compatible. Mixable in the formulation for application together in the same carrier.
Concentration. Amount of active ingredient expressed as percent.
Contact herbicide. Phytotoxic by contact with plant tissue rather by translocation.
Cordate. Hard-shaped with the point away from the base.
Corolla. A collective name for the petals of the flower.
Cotyledon. Primary seed leaf in the embryo.
Creeping. A trailing shoot that produces root in each node.
Culm. Stem of grasses or sedges.
Defoliant. Any substance which causes the leaves or foliage to drop from the plant.
Dehiscent. Opening regularly.
Dioecious. Unisexual with stamnate and pistillate flowers on different plants.
Dormancy. Inhibited germination of seeds or growth of plant organs.

Early post-emergence. After emergence at cotyledonary growth phase of seedlings.

Embryo. New plant in the ovule before germination of the seed.

Emerged plant. A rooted or anchored aquatic plant, with its leaf above water surface.

Emulsion. Suspension of one liquid as minute globules in another liquid.

Endocarp. Inner layer of a pericarp (matured ovary).

Entire. Margin of leaf even, not toothed or divided.

Epicarp. Outermost layer of the fruit.

Exocarp. Outer layer of a periocarp.

Floating plant. A free floating aquatic plant adapted to grow its vegetative tissues above the water surface.

Florate. Individual flower of grasses and compositae.

Foliar application. Application of a herbicide to the leaves or foliage of plants.

Follicle. Fruit consisting of single carpel, splitting along the stem only.

Formulation. A herbicidal preparation or process, supplied by a manufactures for practical use.

Glabrous. Without hair, smooth.

Glume. A small chaff-like bract, those of grass spinkelet.

Granule (gr). A dry formulation of herbicide in minute grains, less than 10 cubic millimeters.

Growth regulator. A substance used for modifying plant growth processes.

Gynoeceum. Assemblage of carpels in a flower with style and stigma.

Herb or Herbaceous plant. A vascular plant but not woody stem above ground.

Herbicide. A chemical used for killing plants, mostly herbs or interrupting their normal growth processes.

Hermaphrodite. A flower in which both stamens and ovary are present and functional.

Hilum. In the seed, the scar or mark indicating the point of attachment.

Imperfect flower. One lacking, either stamens or carpels.

Indehiscent. Not opening by valves or spores.

Inferior ovary. Sepals attached on the top.

Irregular flower. Parts of a whorl or series not all alike.

Lanceolate. Like and lance.

Lateral. Situated to the right or left of medium plane, remote from the parent axis.

Latex. Milky juice.

Legume. Roots from nodules for nitrogen fixation; dehiscent dry fruit splitting along two sides.

Ligule. An outgrowth from top of the sheath in grasses.

Linear. Long and narrow with parallel margins.

Locualr. Compartments in any ovary or fruit just before dehiscence.

Marginal. Outer flowers in the leaf of compositae, either ligulate or tubular.

Monoecious. Staminate and pistillate flowers on the same plant.

Mutagen. A compound having the property to induce mutations.

Naked flower. Without calyx or corolla.

Necrosis. Localised death of living tissue, desiccation and loss of function.

Neutral flower. Without stamens or carpels.

Non-selective herbicide. A chemical generally toxic to plants irrespective of species.

Noxious weed. A weed especially undesirable, troublesome and difficult to control.

Nut. A dry indehiscent fruit, one seeded.

Oblong. Larger than broad.

Obtuse. Blunt but scarcely rounded.

Opposite. Two leaves at a node.

Ovate-lanceolate. Between ovate (broad towards base) and lanceoalte (like a lance).

Ovule. Under-developed seed.

Palmately compound. Leaflets eradiating fan-like from one point.

Panicle. An elongated, irregularly branched, receme like inflorescence.

Parasitic. Growing on and deriving nourishment from other plant.

Pedicel. Stalk of an individual flower in an inflorescence.

Peduncle. Stalk of a solitary flower.

Pellet. A dry formulation of herbicide in descrete particles usually larger than 10 cubic millimeters.

Perennial. Plants that live for more than two years.

Perfect flower. Having both stamens and carpels.

Perianth. Petals and sepals taken together.

Pericarp. Matured ovary.

Persistent. Remaining attached, not all falling off at the same time.

Persistent herbicide. A herbicide which harm susceptible crops planted in normal rotation, after hardship of the treated crop.

Phytotoxic. Injurious or toxic to plants.

Pinnately compound. Leaflets arranged on each side of a common axis of a compound leaf.

Pistil. The ovule bearing organ with style and stigma.

Plumule. Bud or growing point of an embryo plant.

Pollen. Male spore borne by the anther, containing male element.

Post-emergence (Post). After emergence of the specific weed or planted crop.

PP m.w. Parts per million by weight.

Pre-emergence (Pre). Prior to emergence of the specified weed or planted crop.

Pre-planting application. Applied on the soil surface before seeding or transplanting.

Pre-planting soil incorporate (ppi). Applied and tilled into the soil before seeding.

Prickle. A small sharp outgrowth from the bark or epidermis.

Procumbent. Trailing upon the ground without sending roots.

Prostrate. Lying flat on the ground.

Pseudocarp. A fruit with axis, peduncle but not considered to proper fruit eg. pineapple.

Pubescent. Covered with soft short hairs.

Racemose. A type of branched inflorescence in which main axis continues to grow and remain stronger.

Rachis or rhachis. Axis of an inflorescence or a spike.

Radicle. Part of the embryo below the cotyledons.

Rate. Amount of herbicide, acid equivalent applied per unit area.

Receptacle. Enlarged end of the peduncle for which other flower parts are attached.

Reticulate. Veins forming a net work.

Rhizome. Underground horizontal stem.

Rosette. A very short axis bearing a donse cluster of leaves.

Runner. A slender trailing stem, sending roots at each node.

Selective herbicide. A chemical that is more toxic to some plant species than to others.

Serrate. Having sharp teeth pointing forward.

Sessile. Without petiole.

Sheath. A long tubular structure surrounding the stem.

Soil injection. Placement of herbicide beneath the soil surface with an injection blade or tine.

Soluble solid. A dry herbicide formulation that is soluble in the carrier liquid.

Spathe. Bract surrounding a spadix.

Spike. A type of inflorescence with sessile flowers.

Spray drift. Movement of airborne spray from intended area of application.

Staminate. Having stamens and no pistil; male.

Sterile. Infertile, barren.

Stigma. Expanded tip of a style to which pollen adheres.

Stipule. A basal appendage of a petiole.

Stolon. Stem that grows horizontally along the ground surface.

Style. Part of the pistil connecting ovary and stigma.

Submersed plant. An aquatic plant adapted to grow with all of its vegetative tissues below the water surface.

Surfactant. A material which improves the emulsifying, dispersing, spreading, wetting properties of liquids.

Susceptibility. Lack of capacity to tolerate herbicide treatment.

Suspension. Finely divided solid particles dispersed in a solid, liquid or gas.

Tendril. Modified leaf or stem part.

Terminal. At the end of a stem or branch.

Testa. Outer covering of a seed.

Thorn. A modified shoot in the form of a hurd spine.

Terminal. At the end of a stem or branch.

Tolerance. Capacity to withstand herbicide treatment without marked deviation from normal growth or function.

Translocated herbicide. A herbicide that is moved within the plant, mostly in the phloem.

Tuber. A thickened part, usually underground stem.

Turgid. Tense with pressure from within, swollen.

Umbel. A type of inflorescence with umbrella like flower cluster.

Unilocular. An ovary divided by partitions into separate compartments.

Unisexual. Of one sex; staminate or pistillate.

Valve. Units or pieces into which a capsule splits.

Versatile. An anther attached above its base to the tip of filament on which it swings.

Volatile. A herbicide having a sufficiently high vapour pressure.

Weed. A plant growing where it is not desired and interfere with normal activities.

Weed control. The process of killing weeds or economic and other reasons.

Weed eradication. Elimination of all live plant parts and seeds of a weed from a site.

Wetting agent. Substance which serves to reduce interfacial tensions and causes spray solutions to make better contact with treated surfaces.

Wettable power (WP). A finly divided dry herbicide formulation that can be suspended readily in the water.